Environmental Sustainability for Engineers and Applied Scientists

This textbook presents key approaches to understanding issues of sustainability and environmental management, bridging the gap between engineering and environmental science. It begins with the fundamentals of environmental modelling and toxicology, which are then used to discuss qualitative and quantitative risk assessment methods and environmental assessments of products. It discusses how business and government can work towards sustainability, focussing on managerial and legal tools, before considering ethics and how decisions on environmental management can be made. Students will learn quantitative methods while also gaining an understanding of qualitative, legal and ethical aspects of environmental sustainability. Practical applications are included throughout, and there are study questions at the end of each chapter. PowerPoint slides and JPEGs of all the figures in the book are provided online. This is the perfect textbook on environmental studies for engineering and applied science students.

Greg Peters is a professor in the Division of Environmental Systems Analysis at Chalmers University of Technology, Gothenburg, Sweden. Before joining academia, he worked in the urban water sector as a practising environmental engineer. He is the author of many highly cited publications on the quantitative assessment of environmental impacts of water infrastructure, food production, resource consumption and biowaste management. He has served as President of the Australian Life Cycle Assessment Society and as a Lead Author for the Intergovernmental Panel on Climate Change (IPCC).

Magdalena Svanström is a professor in the Division of Environmental Systems Analysis at Chalmers University of Technology, Gothenburg, Sweden. She performs research in environmental and sustainability assessment, in particular in the fields of wastewater and sludge management and biobased production, and also in engineering education for sustainable development. She has been a postdoctoral researcher at Massachusetts Institute of Technology and recently taught life cycle assessment as a visiting professor at Brescia University, Italy. She has published a textbook on sustainable development for engineers in Swedish and has been a member of a United Nations Economic Commission for Europe (UNECE) expert group on educator competences in education for sustainable development.

'This book is a perfect guide for learning how to analyse the problems of natural systems caused by human exploitation, emissions and disturbances. The book analyses the main problems in a scientific way and explains the mathematical models that describe the natural systems. In this way, the book perceives the world from an engineering perspective, but admits that this is a limited perspective, and other disciplines are needed to work towards solutions. It sketches environmental regulation and decision-making. The book is required reading for applied science students and science-driven engineers who are committed to contributing to the grand challenges of our planet.'

— Karel Mulder, The Hague University of Applied Sciences and Delft University of Technology

'Sustainable development – balancing economic and social development with environmental protection – has become a modern paradigm in our technological age. New technological discoveries and product designs are aimed at enhancing human well-being. But they may also give rise to various unwanted "side-effects" that put at risk the quality and longer-term viability of the Earth's biosphere. Such side-effects induce potential environmental hazards on a local, regional and global scale. Engineers are in the vanguard of those using their skills to mitigate these environmental burdens and improve the quality of life. The present contribution, developed by two experts in environmental engineering and management from Chalmers University in Sweden, provides a guide to the tools that the engineer will need in the twenty-first century to combat ecological hazards. They set out the core elements of environmental modelling and toxicology, alongside methods for risk assessment and product environmental life-cycle assessment (LCA). Multi-criteria decision analysis (MCDA) is then advocated as a vehicle for bringing together the quantitative and qualitative appraisal of diverse impacts. Modern concepts like the circular economy (resource efficiency) and the bio-based economy, that may aid industrial decarbonisation, are described. Importantly, the authors also set out the business and government framework for sustainability assessment, including the ethical issues that will necessitate legal tools and management decision-making. This text deserves to be on the bookshelf of all those engineers and applied scientists who will inevitably face the many environmental and sustainability challenges that will confront our planet in the future.'

— Geoffrey Hammond, University of Bath

'Peters and Svanström have produced a valuable textbook for engineers and applied scientists, one which is both comprehensive and most informative across a range of relevant topics associated with environmental sustainability. They manage to convey to the reader the intricate and interconnected quantitative and qualitative aspects of engineering practice as applied in a sustainability-informed context, in a way that can help develop the critical thinking faculties of both students and professional practitioners. In short, *Environmental Sustainability for Engineers and Applied Scientists* represents a valuable and unique addition to the field, not just in relation to (engineering) education for sustainable development, but as a support for best-practice contemporary engineering education more generally.'

— Edmond Byrne, University College Cork

'A comprehensive textbook with a content of key knowledge essential to understanding sustainability and environmental management. The book will help to bridge the gap between engineering and environmental science. I would strongly recommend the book for both undergraduate and master's students at technical universities and faculties.'

— Fredrik Gröndahl, Royal Institute of Technology, Stockholm

Environmental Sustainability for Engineers and Applied Scientists

GREG PETERS

Chalmers University of Technology, Gothenburg

MAGDALENA SVANSTRÖM

Chalmers University of Technology, Gothenburg

CAMBRIDGE
UNIVERSITY PRESS

CAMBRIDGE
UNIVERSITY PRESS

University Printing House, Cambridge CB2 8BS, United Kingdom

One Liberty Plaza, 20th Floor, New York, NY 10006, USA

477 Williamstown Road, Port Melbourne, VIC 3207, Australia

314–321, 3rd Floor, Plot 3, Splendor Forum, Jasola District Centre, New Delhi – 110025, India

79 Anson Road, #06–04/06, Singapore 079906

Cambridge University Press is part of the University of Cambridge.

It furthers the University's mission by disseminating knowledge in the pursuit of
education, learning, and research at the highest international levels of excellence.

www.cambridge.org
Information on this title: www.cambridge.org/9781107166820
DOI: 10.1017/9781316711408

© Greg Peters and Magdalena Svanström 2019

First published 2019

Printed in the United Kingdom by TJ International Ltd. Padstow Cornwall

A catalogue record for this publication is available from the British Library.

ISBN 978-1-107-16682-0 Hardback
ISBN 978-1-316-61773-1 Paperback

Additional resources for this publication at www.cambridge.org/eseas

Contents

Preface

The first question a potential reader might ask is: 'Why this book?' Before deciding to make the effort of writing this book, we had to ask ourselves the same question. Our short and surprising answer was: we looked but we could not find a suitable alternative. We meet hundreds of students each year and were challenged to find them a book that is both quantitative and holistic. We wanted a book that built on the basic mathematics and chemistry that most engineers and applied scientists have by their second year of university. (Rigorous high schools may take students to the same level in some countries.) But we expect our students to develop a holistic perspective on environmental management, so we also wanted a book that was broad in the sense that it includes but also goes beyond environmental science, to encompass environmental information tools, ecological issues, law, decision-making and ethics. After fruitful discussions with Cambridge University Press and useful feedback from some anonymous reviewers of our initial plan, this book is the result.

This is not a book on technical design of engineering interventions (air filtration systems, wastewater treatment plants and so on). Plenty of other books are available to introduce environmental technology solutions. Instead, the intent here is to introduce the reader to ways in which we can scientifically describe the environmental problems that need solving, to help the reader to understand some of the management tools that exist for companies and governments that want to solve them, and to choose between alternative solutions. Of course, 'sustainability' is a very broad word, and a wide variety of professionals are working towards it. This book is not intended to educate biologists, ecologists, sociologists, political scientists or other professionals in their core skillsets, but we acknowledge that professionals like these also have important contributions to make to environmental management. We hope this book will help engineers and other applied scientists to broaden their core skills and communicate with people of other professions.

We would like to acknowledge some of the many people who have helped to make this book possible; educators who have collaborated with us and inspired this work include Sara Heimersson, Hanna Holmquist, Stuart Khan, Stephen Moore, Kathleen Murphy, Gustav Sandin Albertsson and Peter Thor. Robin Harder deserves particular thanks for his collaboration with Greg Peters on teaching materials to the Chemical Environmental Science course at Chalmers University of Technology, which accelerated the development of Chapter 4. Chalmers' provision of his time and partial funding for us is gratefully acknowledged.

Reader's Guide: How Is All of This Connected?

Environmental problems can be viewed from many perspectives, and solutions can be imagined at many levels of detail. It can be hard to visualise how different perspectives fit together. So, to help you understand this book, this reader's guide provides some examples of how different chapters of this book relate to each other. Notice the arrows in the figure indicating cause-effect and informational connections between topics mentioned in the chapters – each of them is briefly explained in the chapter narrative below it.

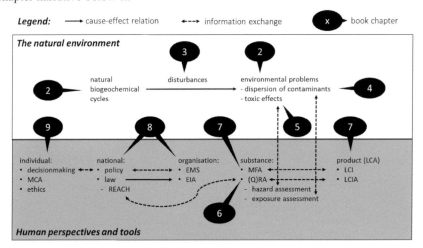

Chapter 1 provides a general introduction to the book by describing some of the overarching concepts that allow engineers to engage with environmental sustainability from an engineering perspective. An understanding of ecological principles and natural biogeochemical cycles is also foundational because many of the problems we face are caused by natural processes that we have perverted: the natural greenhouse effect is necessary for our survival, but we have intensified it; natural transport and reactive processes in ecosystems can dilute and destroy pollution, but they can also contaminate pristine locations or even make some toxins more potent. So Chapters 2 and 3 describe the natural world and our impacts on biogeochemical cycles. To help to explain how pollution spreads, an introduction to modelling the dispersion of contaminants in the environment is provided in Chapter 4, along with mathematical examples. This is complemented by a more qualitative introduction to toxicology which describes how contaminants are transported in, transformed by and impact on the human body (Chapter 5). These two chapters are critical to understanding the one that follows, because predicting contaminant concentrations in the environment and describing their effects on people and other organisms are critical steps in the process of (quantitative) risk assessment (QRA) (Chapter 6). Risk assessment is a key tool that applied scientists and engineers often use to provide a perspective on the risk posed by substances, but Chapter 6 also introduces qualitative risk assessment methods in relation to the need to manage risk in factory operations.

Another substance-level perspective is offered by material flow analysis (MFA) in Chapter 7, a chapter which mostly takes a product-level perspective of environmental management. An MFA can be thought of as a map of substance flows for a region, whereas producing a life cycle inventory (LCI) is more typically about mapping the supply chain for a product. Some of the same data are used in compiling MFAs and LCIs. LCI is the second step in life cycle assessment (LCA), where the third is life cycle impact assessment (LCIA). LCIA methods for chemical impacts are based on QRA methods. The EU REACH legislation is a good example of risk assessment embodied in legislation, which is one of the organisational and national tools discussed in Chapter 8. If you are ever involved in writing or reviewing the environmental impact assessment (EIA) process for something an organisation wants to build, then the general structure of that EIA will be predefined in the local laws – a legal cause-effect relationship. Many private companies have an environmental management system (EMS), but EMSs are also employed by national governments, among others, and it can be said that Sweden's Environmental Objectives are part of an EMS policy operating at the national level. Ultimately, policy decisions are made by people. The final chapter of the book takes the perspective of people working in organisations and us as individuals, and examines factors that influence decision-making. It also suggests how we can balance environmental and other goals in practice by considering mathematical support offered by multicritera analysis (MCA) and more qualitative ethical principles (Chapter 9).

There is more to this book than just these topics, but hopefully you now have a sense of how some of them are related to each other in the real world. Each chapter comes with some study questions at the end to help you to process what you have read and to extend your understanding into new places and applications not directly covered here.

Abbreviations

AHP	analytic hierarchy process
BAT	best available technology
BATNEEC	best available techniques not entailing excessive costs
BPEO	best practicable environmental option
BSL	biosafety level
CFC	chlorofluorocarbon
CLRTAP	Convention on Long-Range Transboundary Air Pollution
CMR	carcinogenic, mutagenic, toxic to reproduction
DPSIR	drivers, pressures, states, impacts, responses
DDT	dichlorodiphenyltrichloroethane
DNA	deoxyribonucleic acid
EChA	European Chemicals Agency
ED_{50}	effective dose for 50% of organisms
EDC	endocrine-disrupting chemical
EEIOA	environmentally extended input-output analysis
EFA	ecological footprint analysis
EIA	environmental impact assessment
EIS	environmental impact statement
EMS	environmental management system
ESA	environmental systems analysis
GHS	globally harmonised system
H	Henry's law coefficient
HAZOP	hazard and operability study
HCFC	hydrochlorofluorocarbon
HQ	hazard quotient
IOA	input-output analysis
K_{oc}	soil organic carbon sorption coefficient
K_{ow}	octanol-water partition coefficient
K_S	solubility constant
LCC	life cycle costing
LCI	life cycle inventory
LCIA	life cycle impact assessment
LCSA	life cycle sustainability assessment
LD_{50}	lethal dose for 50% of organisms
MAUT	multiattribute utility theory

MCA	multicriteria analysis
MCDA	multicriteria decision aid
MFA	material flux (or flow) analysis
MIPS	material intensity per service
NGO	non-governmental organisation
NIMBY	not in my backyard
NOAEL	no observable adverse effect level
NPV	net present value
PBT	persistent, bioaccumulative and toxic
PCB	polychlorinated biphenyl compound
PDCA cycle	plan-do-check-act cycle
PER	public environmental report
PET	polyethylene terephthalate
POP	persistent organic pollutant
PROMETHEE	preference ranking organisation method for enriched evaluation
QALY	quality adjusted life years
QRA	quantitative risk assessment
REACH	Registration, Evaluation and Authorisation of Chemicals Regulation
RfD	reference dose
RoHS	Restriction of Hazardous Substances Directive
SLCA	social life cycle assessment
SVHC	substances of very high concern
UF	uncertainty factor
UNFCCC	United Nations Framework Convention on Climate Change
USEPA	United States Environmental Protection Agency
vPvB	very persistent and very bioaccumulative

The Engineer's Role in Environmental Protection

Our relationships with nature are complicated. On the one hand, we depend completely on functioning ecosystems for the provision of food and many other services, but on the other hand, our activities have impacts on natural systems, threatening the long-term prosperity of human society and the survival of other species. Engineers are key facilitators in these relationships, mechanising and optimising the flows of valuable resources into human society, and accelerating or moderating the damage done to nature. This realisation has driven various attempts by engineers and other professionals to employ sustainable development in practice.

In order for the urgent transition to sustainability to take place, all business activities, government policies and industrial projects need to be considered in terms of their potential impacts on different areas of the economy, society and nature. Very often, trade-offs will have to be managed, and this frequently involves negotiation between stakeholders, each with its own set of experiences, interests and values. In this context, what is the role of the engineer or applied scientist? This book deals with this question and provides some knowledge and tools. As we have worked with the water, materials, textiles, food, chemicals, waste management and regulatory sectors of the economy, much of this book is based on the needs of engineers and applied scientists (we use these terms interchangeably) who are employed in these sectors, but it will be relevant also to other engineers facing environmental management challenges connected with resource use and pollution.

1.1 The Concept of Sustainable Development

Since the 1970s, the concept of sustainable development has increasingly been used to describe the need to find new societal models that could keep generating economic growth, and thereby continue to increase welfare levels in the world while also protecting the environment. Some particular environmental issues became critical then because the size of human settlements and the scale of industrial production had developed dramatically over the previous 100 years. In particular, acid rain and persistent toxic pollutants were discussed at the time. In Scandinavia, for example, fish were dying in acidified lakes. The acidification was caused by emissions originating in other countries where coal burning was common. Meanwhile, predatory birds and marine mammals in the Great Lakes of North America and in the Baltic Sea in Northern Europe were suffering reproductive problems on account of persistent chemicals in their food chains. It was clear that environmental problems could not be solved on a national basis and that

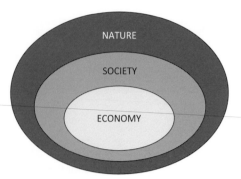

Figure 1.1 Sustainable development described as the consideration of three areas but where economy operates within society which in turn operates within the environment (nested dependencies model; after Doppelt, 2008), in contrast to models that separate the three areas and place them as partly overlapping circles

the global community needed to come together to create a common vision and find ways to cooperate to establish a more 'sustainable' kind of development.

The definition of sustainable development provided by the Brundtland Commission in its report 'Our Common Future' in 1987 is the most commonly quoted: 'development that meets the needs of the present without compromising the ability of future generations to meet their own needs'. At the heart of this definition is the fulfilment of human needs, now and in the future. However, we may also choose to introduce other values into this concept, such as the right of other species to thrive as well. In fact, discussions on sustainable development are often criticised as too anthropocentric, particularly by people with more Earth- or life-centred worldviews who see an inherent value in nature beyond its value in directly serving mankind via the provision of ecosystem services (discussed in Chapter 2). For more discussion of other worldviews, see Chapter 9. Common to different definitions of sustainability or sustainable development is that they all mention diverse dimensions of the world that need to be viewed simultaneously in a holistic way. With this perspective, different positive and negative impacts, as well as potential trade-offs, can become visible. Most commonly, we express this as a need to simultaneously consider environmental impacts, economic impacts and social impacts, and illustrate it by means of three overlapping circles or three pillars. However, more or fewer dimensions and other terms can occur in other conceptualisations of sustainable development. When the hard limits posed by the function of natural systems are emphasised over the softer limits of the man-made economic system, it is common to place economy and society within the realm of nature, as illustrated by the nested dependencies model in Figure 1.1 (see Doppelt, 2008). Today, the global dialogue on sustainable development and what it means in practice is perhaps best described by the 17 sustainable development goals; see Table 1.1.

1.2 Unique Skillsets and Perspectives of the Engineer

'Engineer' is an ancient word which acquired some qualifying prefixes in the 18th and 19th centuries, creating the branch terms we recognise today. Chemical

Table 1.1 The United Nations Sustainable Development Goals	
Goal	Statement
1	End poverty in all its forms everywhere
2	End hunger, achieve food security and improved nutrition and promote sustainable agriculture
3	Ensure healthy lives and promote well-being for all at all ages
4	Ensure inclusive and equitable quality education and promote lifelong learning opportunities for all
5	Achieve gender equality and empower all women and girls
6	Ensure availability and sustainable management of water and sanitation for all
7	Ensure access to affordable, reliable, sustainable and modern energy for all
8	Promote sustained, inclusive and sustainable economic growth, full and productive employment and decent work for all
9	Build resilient infrastructure, promote inclusive and sustainable industrialization and foster innovation
10	Reduce inequality within and among countries
11	Make cities and human settlements inclusive, safe, resilient and sustainable
12	Ensure sustainable consumption and production patterns
13	Take urgent action to combat climate change and its impacts
14	Conserve and sustainably use the oceans, seas and marine resources for sustainable development
15	Protect, restore and promote sustainable use of terrestrial ecosystems, sustainably manage forests, combat desertification, and halt and reverse land degradation and halt biodiversity loss
16	Promote peaceful and inclusive societies for sustainable development, provide access to justice for all and build effective, accountable and inclusive institutions at all levels
17	Strengthen the means of implementation and revitalize the global partnership for sustainable development

engineering deals in particular with the application of chemistry, mathematics, physics and biology to perform chemical processes on a commercial scale. Civil engineering is also a broad field but more concerned with applying mathematics, physics and material and environmental science to the development of infrastructure, including roads and buildings. In mechanical engineering, engineers typically use similar scientific concepts and thermodynamics in the design of automated mechanical systems. There are many other branches of engineering (electrical and electronic engineers are the most numerous), not to mention specialisations within the fields mentioned here, but the point is that they all use natural scientific knowledge and mathematics in their professions in a way that many other professions do not. Sometimes engineering is referred to as applied science, and it is therefore natural that the types of knowledge and paradigms that exist in science also exist in engineering, although complemented with a more pragmatic ontological stance as engineering is not only about the search for what is 'true' but also for what is effective and useful. Furthermore, engineering is not only about the function of the technology itself but also its use, management, policy, economy and various other aspects of technology.

What is the role of engineers in the context of *sustainable development*? What specific skillset and perspectives does engineering provide? How is this useful in complementing other skillsets and perspectives? What are the limitations of the skillsets and perspectives of engineers and what are the dangers of applying only an engineering perspective to problem-solving? This chapter explores these issues and provides some mental models and tools that are particularly useful in different engineering contexts – they were generated using an engineering perspective. Consequently, they may not represent and provide all the necessary information for different projects and must therefore be complemented with models and tools that represent other perspectives.

One important way in which engineering differs from many other fields is that it often deals with the quantitative modelling of various technologies. Engineers are often very skilled in mathematics and, based on analysis of various quantitative parameters of a technology or a technical system, engineers can often model and predict the expected behaviour. Systems analysis is a set of tools that are common in engineering and engineers often use systems analysis tools combined with their mathematical skills to describe fairly complex technical systems, such as factories, cars or computers. Predicting how complex systems will operate and the environmental impacts they will have is one of the important contributions of engineering science to sustainable development – without it policy makers may at worst be left with glib, simplistic ideas about sustainability, such as that more recycling is always better. An engineer can identify scenarios under which the energy used and pollution generated by a recycling system exceed the energy used and pollution generated by using raw materials to make the same product. Engineers can introduce quantitative information which would otherwise be missing in planning and policy debates.

When it comes to even more complex sociotechnical systems such as the behaviour of traffic or food provision for a whole city, engineering methods may be inadequate and only useful for some aspects and under certain conditions. The more complex the modelled system is, the less the modelling will represent reality, as uncertainties and complex interactions come into play. There is a danger in expecting that all important aspects of a system can be described in mathematical terms and its behaviour thereby predicted. In particular, when the studied systems are not purely technical but rather sociotechnical or even sociotechnoecological, such as for managing climate change, problem-solving becomes challenging and such problems are often referred to as *messy, ill-structured* or *wicked*, and they require other management approaches to complement the traditional engineering methods. Some approaches include attempts to use different perspectives in the same project (multidisciplinarity), to create new perspectives in which elements from several others are present (interdisciplinarity) or even go beyond disciplines to work together with society in problem formulation and attempting to find workable measures that will lessen a complex (and possibly even wicked) problem and lead to an overall improvement (transdisciplinarity). In some contexts, 'narrative approaches' that embrace qualitative and path-dependent information (Andersson et al., 2014; Boje, 2001) will have to be applied for describing the functioning of wicked systems.

Inherent in the tools that are predominant in engineering are certain epistemological and ontological perspectives. For example, *positivism* is a research philosophy that sees factual knowledge based on observation, including measurement, as trustworthy. The role of the researcher is to collect data and to use as objective an approach as possible in interpreting them. Positivism is based on an atomistic view of the world, which means that it assumes that the world can be divided into discrete, observable elements and events that interact in a determined and regular manner. Results from a study based on a positivist philosophy are typically observable and quantifiable and experiments can be copied and will generate similar results, providing managers and decision-makers with similar recommendations. As the saying goes, 'you cannot manage what you cannot measure.' But it goes without saying that there are systems and phenomena that may be relevant to the broad field of engineering that such approaches might fail to capture.

This is not to go to the extreme of proposing that facts do not exist outside of an observer or that the concept of sustainability is too wicked for use. Some facts regarding the relationships between systems can be established and measurable aspects of environmental systems can be considered *absolute* preconditions for sustainability (for example, where climate change results in a local wet-bulb temperature of 35°C, mammals will die of heat stress because dissipation of metabolic heat becomes thermodynamically impossible; see Sherwood and Huber, 2010). So there is great value in the work which engineers do to describe an environmental or technical process in mathematical terms, and to calculate optimal process settings for the variables we control, but there is a limit to what can be described in such terms. In short, there is sometimes a perception in science and in engineering of a dichotomy between quantitative and qualitative research when in fact, these are complementary and will answer different questions.

1.3 Sustainability as an Engineering Problem

Technological systems often require energy and materials both for the manufacturing of their technical components and for their operation. Therefore, the design and practical use of technology will often determine the energy and material throughput of modern society. Unfortunately, these energy and material flows can lead to severe problems of resource availability and pollution. The actions of engineers will therefore have a strong impact not only on the well-being of people whose needs they aim to fulfil but also on the environment. An engineer therefore needs a systems perspective to be able to understand the impacts of technology, for example from a life cycle perspective. Along the value chains that are generated as different industrial and societal actors get involved in the life cycle of a certain technology, there may also be social impacts on people that are affected, for example fashion retailers in Sweden influence workers in a Bangladeshi clothing factory, and electronics manufacturers in Asia influence workers in a mine in South Africa. Therefore, a holistic approach is also needed that considers various types of impacts in an integrated manner. This systems view is needed both for assessing current technologies, for understanding what needs to change and for understanding the potential impact of new designs and of scale-up of new technologies.

Within the broad framework of industrial ecology, there are various informational tools for assessing and optimising stocks and flows of material and energy resources. In this chapter of the book we focus on the broad framework and some examples of principles and systems-based models that engineers may employ. In later chapters, some of the information tools of industrial ecology are described, such as life cycle assessment and material flow analysis (Chapter 7) and multicriteria analysis (Chapter 9).

1.3.1 Industrial Ecology

'Industrial ecology' was first mentioned within the field of economic geography in the late 1940s, but has now become an established field of study, manifested primarily within engineering sciences. So although the term appeared earlier, it is generally said to have been coined in 1989 by Frosch and Gallopoulos (1989).

The term *industrial ecology* makes a metaphor between technological and natural systems, suggesting we need to mimic how energy and material flows are handled in nature, using basic laws of physical chemistry and thermodynamics in the analysis of energy and material flows. In an ecosystem, species feed on each other's wastes. Technological systems should thus be organised so as to facilitate circular flows of materials (Graedel, 1994). Circularity is thus a key concept in industrial ecology. Further, most of the energy input to ecosystems is solar energy, and when the 'producers' (the green plants) have tied up some of this energy in biomass, it can be accessed by 'consumers' and 'decomposers'. Thus, reliance on solar-based renewable energy is another key concept within industrial ecology. Another key aspect of ecosystems is the interaction between different types of organisms, with diversity as an important feature providing resilience (the ability to cope with change) to ecosystems.

When principles of industrial ecology are applied to industrial activities, in particular when it deals with the exchange of energy and material flows between different industrial facilities, this is often referred to as *industrial symbiosis*, another metaphor from ecology.

Inspired by the framework of industrial ecology and the traditional perspective of an engineer, several different sets of sustainability principles or sustainability criteria have been proposed. Some of these are described in what follows. There is a strong focus on energy and material flows caused by the production and use of technology in these principles and criteria, so other aspects may be missing. This is something that engineers need to pay attention to so that the whole breadth of relevant sustainability aspects is considered in practice.

1.3.2 Overall Sustainability Principles for Activities in the Earth System

1.3.2.1 Circular or Biobased Economy

As long-term visions and principles for the establishment of economic activities in society, the concepts of *circular economy* and *biobased economy* have become strong in the past 20 years, and they have been the basis for both development of national policies and other strategies and principles. Both circular and biobased economy are manifestations of industrial ecology thinking on a societal level. The two concepts often strongly overlap but also emphasise different aspects of energy and material flows.

Circular economy focusses primarily on the circulation of material flows and suggests that the economy should rely on activities that enable such circularity. In China's 11th five-year plan, starting in 2006, promoting a circular economy was identified as a national policy. A circular economy is the basis for the cradle-to-cradle framework developed by Braungart and McDonough. In 1995, they started a consultancy and later published a book that described their framework (McDonough & Braungart, 2002). Their framework emphasises the need for circularity and they suggest that all flows in society need to be designed to be part of either technical or biological recycling systems so that the flows can be seen as either technical nutrients or biological nutrients. They also stress that nature is not always efficient but it is effective, and our current focus on improving efficiency should be shifted to or at least complemented with a focus on effectiveness, that is, on doing the right things. They argue that if we are doing the wrong things, it does not matter how efficient we are in doing them.

Starting in 2010, the Ellen MacArthur Foundation picked up these ideas and developed them further. The Foundation's framework for circular economy (www .ellenmacarthurfoundation.org) is illustrated in Figure 1.2. The different strategies shown in Figure 1.2 speak clearly to industrial actors on what options they have to closing loops. The flows of renewables to the left in Figure 1.2 are the flows that McDonough and Braungart call biological nutrients, and the stock flows to the right are the technical nutrients.

In practice, the pursuit of a circular economy today often results in various reuse or recycling efforts and a rethinking of product design to allow for reuse or recycling. It is often emphasised in these types of frameworks that the best form of recycling is closed-loop recycling that avoids the down-cycling to products of lower quality, as this only delays disposal rather than provides true circularity to the societal metabolism.

The biobased economy (or bioeconomy) instead focusses on the need to move towards biobased production. This is a response to our dependence on finite fossil resources and the problem of climate change that results from the combustion of products and fuels made from them. The biobased economy is of course particularly attractive to countries and companies with considerable biomass resources (e.g. forests). Market interest in paper products is decreasing. So in Sweden and Finland, for example, the biobased economy is seen as an opportunity to strengthen the forest industry by means of value-added products from new technologies being introduced in existing pulp mills, or from new biorefineries using new combinations of technologies to produce a variety of products from wood or other biomass. In 2012, former US president Barack Obama announced a National Bioeconomy Blueprint, describing the bioeconomy as 'economic activity powered by research and innovation in the biosciences'.

The biobased economy is not necessarily something completely different to a circular economy. Going back to Figure 1.2, the biobased economy can be seen

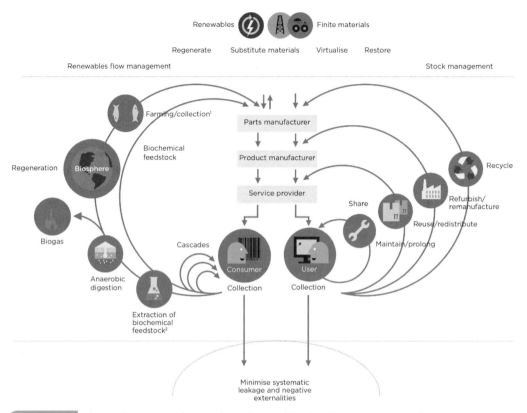

Figure 1.2 The circular economy framework as illustrated by the Ellen MacArthur Foundation (www.ellenmacarthurfoundation.org). Adapted with permission from the Ellen MacArthur Foundation

as the left-hand side that deals with biological nutrients and particularly in the sense that it relies on nature's ability to circulate carbon flows and nutrients such as nitrogen and phosphorus.

1.3.2.2 Four Principles for Sustainability

In the 1990s, a set of four principles (or 'system conditions') for sustainability was developed by the Swedish researchers John Holmberg, Karl-Henrik Robèrt and Karl-Erik Eriksson (Holmberg et al., 1995; Robèrt et al., 1997). They suggested there were basic scientific foundations on which such principles could be developed. In particular:

- Nothing disappears; mass and energy in the universe are conserved; mass and energy may be converted into different forms, but the total amount of energy in an isolated system remains constant (the principle of matter conservation and the first law of thermodynamics).
- Everything spreads; energy and matter tend to dissipate spontaneously (the second law of thermodynamics, or the law of entropy).

- There is value in structure; material quality lies in concentration and structure of the matter that makes up a material; what we consume are the qualities of matter rather than matter itself.
- Biological plants create structure and order by using energy from the sun; net increases in material quality on Earth are generated almost entirely by photosynthesis.

On the basis of these foundations, they proposed four principles. The original principles are reproduced here (with some explanatory parenthetics). In a sustainable society:

(1) Substances extracted from the lithosphere must not systematically accumulate in the ecosphere (for example fossil CO_2 and heavy metals).
(2) Society-produced substances must not systematically accumulate in the ecosphere (for example antibiotics and endocrine disruptors).
(3) The physical conditions for production and diversity within the ecosphere must not be systematically deteriorated (for example by deforestation and draining of groundwater tables).
(4) The use of resources must be effective and just with respect to meeting human needs.

The first three principles are inspired by a systems analysis of the Earth system, using the basic laws of energy and matter. Therefore, they are aligned with industrial ecology thinking and represent an engineering view of the world. The fourth principle, shown here in its original wording, was intended to reflect an engineering perspective on the existence of limits to global resources, but also the need to share these resources in order to fulfil normative goals concerning human needs and intra-generational equity within the sustainable development discourse.

The Swedish non-governmental organisation (NGO) The Natural Step (TNS) took on the role of promoting these principles internationally and today works with organisations all over the world to try to implement them in corporate practice. TNS presents these principles as a set of rules that any government or business activity will have to comply with in order to have a place in a sustainable world, thus making them a useful set of criteria for evaluating long-term strategic planning. TNS has also reworded the principles several times (for example using the grammatically better 'degraded' instead of 'deteriorated' in the third principle). Of the four, the last principle has been the subject of the most dispute and revision, shifting it further from the scientific towards the humanistic domain. The latest wording at the point of publication of this book says a sustainable society has 'no structural obstacles to people's health, influence, competence, impartiality and meaning' (TNS, 2018).

1.3.3 Systems-Based Models Describing Causes or Cause-Effect Relationships

1.3.3.1 IMUP Equation

The environmental problems that became increasingly pressing in the 1970s were a direct result of the increasing production and consumption in society that was made possible by the technological and societal development we sometimes call *industrialisation*. The so-called IMUP equation illustrates the connection between industrialisation and environmental problems (Azar et al., 2002). It is a development of the so-called IPAT equation that Ehrlich and Holdren proposed in 1970 (Chertow, 2001). The IMUP equation reads:

$$I = i * m * u * p \qquad\qquad (1.1)$$

Where I is the total impact in some kind of quantitative impact unit, p is the human population of the considered region, u is the utility level for the region given in the number of 'functions' that are provided to each person (e.g. the number of km transported or kWh used per capita), m is the material and energy intensity given as mass and energy units per provided function, and i is the impact per mass and energy unit. (In the predecessor equation called IPAT, the i and m factors were combined into one technology factor.)

This decomposition of the total impact into four factors is an engineering model for illustrating the sustainability challenges and some available strategies. With a simple calculation example, this equation can help us to understand the magnitude of the challenges that we face. Many different assessments (see further descriptions in Section 2.5) indicate that the pressure on the environment needs to be greatly reduced. Depending on the type of impact, we might come up with many different estimates for how much the total impact needs to decrease, but let us consider the 2017 estimate by the Global Footprint Network of how much the global ecological footprint exceeds the carrying capacity of the Earth, in order to come up with a number. They estimate that we are using an area equivalent to the available area on 1.6 Earths. The total impact therefore needs to be reduced by almost 40% in order to prevent our consumption from exceeding the Earth's carrying capacity. At the same time, population is increasing. It is currently estimated that in 2050, the world population will be around 9.7 billion, an increase by about 1.9 billion compared to 2017, or almost 30%. We can also expect that consumption levels will increase as this is a global trend, and in particular in poor regions as these are experiencing economic growth. Let us assume that the global consumption level will triple until 2050, which is probably a low estimate, thus the utility level will increase to 300% of the present. If we are going to reduce the total impact to 2050 to the level stated earlier, with a simultaneous increase in global population and in consumption levels, we can calculate the need for changes in the i and m factors (the technology factors in the earlier IPAT equation) that are related to how much and what kind of material and energy we use for providing the utilities, i.e. changes that we need to make in industry in terms of efficiency, raw material base etcetera. The calculation shows that these two factors together need to be reduced to only about 16% of what they are today, i.e. the *eco-efficiency* needs to increase by more than a factor of 6 (see Chertow, 2001, for a description of the *Factor X* concept). The term *eco-efficiency* is commonly used also in other contexts to describe how many goods or how much value we can create out of each unit of resource or per environmental impact unit. Remember that the calcula-tion presented earlier represents a low estimate for the challenges that lie ahead; various organisations have suggested factors between 4 and 50.

The IMUP equation also implicitly suggests the different possibilities we have for reducing the pressure on the environment. Looking at all four factors, we can reduce

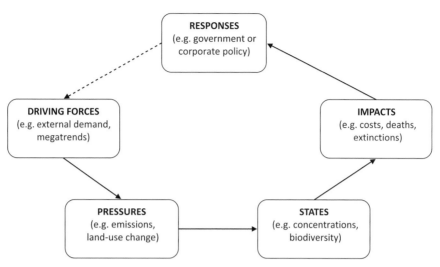

Figure 1.3   The DPSIR framework

population, reduce the number or type of utilities that each person uses, reduce energy and material use per utility (*dematerialisation*) or change to other types of energy and material (*transmaterialisation*). Within each of these four areas, there are, in turn, a number of more detailed strategies to employ, and many of them can of course be combined. To connect this to some concepts discussed earlier, dematerialisation efforts often refer to the same types of activities as are emphasised in a circular economy: reduce, reuse and recycle. Transmaterialisation, however, is more closely connected to e.g. bioeconomy efforts as the idea is that it will create less impact to shift to biobased production.

The IMUP equation therefore shows us what the important factors are behind the total impact and points to what different types of strategies we can use to reduce the impact. But there are also systems-based models that help us focus more on cause-effect relationships than the causes themselves. One such example is the DPSIR framework.

1.3.3.2 DPSIR Framework

The DPSIR framework was developed in the 1990s (Smeets & Weterings, 1999) and has been adopted by the European Environment Agency (EEA). The letters in the name (DPSIR) refer to the different steps in the cause-effect chain from drivers to response; see Figure 1.3. To illustrate using climate impacts of the energy system, a *driving force* is e.g. the increasing energy demand, a related *pressure* is carbon dioxide emissions from burning of fossil fuels, a related *state* in the environment is carbon dioxide concentration in the atmosphere, a related *impact* is a climate change, e.g. in terms of wilder weather and changing precipitation patterns, and related *responses* can be e.g. an energy tax that will reduce the driving force of consumer demand, or a carbon dioxide emission tax that will reduce the pressure by causing a technological shift towards renewables.

In 2004, the EEA developed a core set of 37 indicators that make DPSIR an analytical framework for informing policy implementation, e.g. in designing assessment of alternative policies, in selecting indicators and in communicating results (EEA, 2005). The number of core indicators was later increased to 42 (EEA, 2014).

Selecting indicators is an important part of any assessment of policies for sustainable development or of the extent to which their implementation was successful. The DPSIR framework is just one of the contexts where the need for relevant and purposeful indicators is obvious. One context where the choice of indicator is critical is in relation to the *planetary boundaries* framework that is described in more detail in Section 2.5. In that framework, researchers have selected certain control variables that describe the state of the environment within nine areas that together provide an understanding of the functioning of natural systems. They acknowledge the lack of knowledge in some areas and even refrain from generating control variables for some areas before more is known. Another context is with regard to the 17 sustainable development goals (see Table 1.1), each of which has several subgoals expressed as indicators that can be monitored. The selection of indicators is a key step in any *multicriteria analysis*, which is a process that seeks to create an overall assessment of alternative scenarios based on multiple criteria. Generic principles for indicator selection are discussed from a more mathematical point of view in Chapter 9.

1.3.4 Practical Design Principles for Industrial Activities

1.3.4.1 Principles of Green Engineering

Based on an understanding of overall principles for the Earth system, more practical principles or sets of principles have been developed for different, more specific contexts. One such example is the set of 12 *green chemistry principles* that was originally developed by Warner and Anastas in the 1990s (Anastas & Warner, 2000). In the following years, several other researchers responded and published new sets of principles for green chemistry or green engineering that were either alternative or complementary (e.g. Anastas & Zimmerman, 2003; Winterton, 2001). These principles are formulated as rules of thumb to be considered in designing chemical processes. The green engineering principles by Anastas and Zimmerman (2003), as provided by the American Chemical Society (ACS, 2018), are:

1. Inherent Rather Than Circumstantial: Designers need to strive to ensure that all materials and energy inputs and outputs are as inherently nonhazardous as possible.
2. Prevention Instead of Treatment: It is better to prevent waste than to treat or clean up waste after it is formed.
3. Design for Separation: Separation and purification operations should be designed to minimize energy consumption and materials use.
4. Maximize Efficiency: Products, processes, and systems should be designed to maximize mass, energy, space, and time efficiency.
5. Output-Pulled Versus Input-Pushed: Products, processes, and systems should be 'output pulled' rather than 'input pushed' through the use of energy and materials.

6. Conserve Complexity: Embedded entropy and complexity must be viewed as an investment when making design choices on recycle, reuse, or beneficial disposition.

7. Durability Rather Than Immortality: Targeted durability, not immortality, should be a design goal.

8. Meet Need, Minimize Excess: Design for unnecessary capacity or capability (e.g., 'one size fits all') solutions should be considered a design flaw.

9. Minimize Material Diversity: Material diversity in multicomponent products should be minimized to promote disassembly and value retention.

10. Integrate Material and Energy Flows: Design of products, processes, and systems must include integration and interconnectivity with available energy and materials flows.

11. Design for Commercial 'Afterlife': Products, processes, and systems should be designed for performance in a commercial "afterlife."

12. Renewable Rather Than Depleting: Material and energy inputs should be renewable rather than depleting.

These principles can actually apply to many kinds of engineering designs outside chemical engineering. On the other hand, the preceding green chemistry principles are naturally more branch-specific, focussing on aspects that are important for chemical synthesis or chemical processing (e.g. 'catalytic reagents are superior to stoichiometric reagents').

1.3.4.2 Waste Hierarchy

Another set of principles to consider in design of industrial activities is implicit in the *waste hierarchy*. It was introduced into European waste policy in the European Union's Waste Framework Directive (1975/442/EEC) as early as 1975. It ranks different waste management strategies from the most favoured to the least favoured, based on typical life cycle environmental impacts and on circular thinking; see Figure 1.4. This hierarchy is thus based on the same industrial ecology thinking as other sets of principles that have been discussed in this chapter, but only provides advice for a specific context. Waste prevention is a principle that is also very strong in the context of initiatives for industrial symbiosis, circular economy, dematerialisation and green engineering.

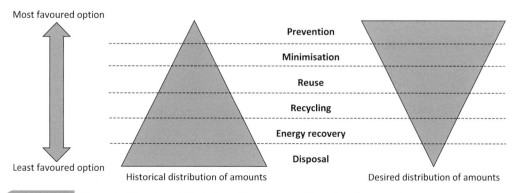

Figure 1.4   The waste hierarchy and illustration of how the pyramid needs to be inverted

The pyramidal shape to the left in Figure 1.4 indicates the general, global situation today in terms of the scale of material flows dealt with (e.g. more waste is disposed to landfill than recycled). The goal is to turn the pyramid upside down, with most of the waste being prevented and the least amount being disposed in landfills; see the right part of Figure 1.4. In practice, the waste hierarchy cannot always be followed slavishly because of other requirements, e.g. hygiene criteria for food packaging.

1.3.4.3 Other Design Principles

Waste prevention and pollution prevention are closely related. After all, pollution is often just waste in liquid or gaseous form. Waste and pollution prevention or abatement are always done in a reality where various other kinds of restrictions may limit what can be done, for example the availability of technology and the economic costs of different methods. Therefore, in law, a compromise concept is often used that has been variously called BAT (Best Available Technology), BATNEEC (Best Available Techniques Not Entailing Excessive Costs) or BPEO (Best Practicable Environmental Option). The concept of BAT was introduced into EU legislation as early as 1984 and today BAT for a given industrial sector is described in BAT reference documents (BREFs) as defined in the Industrial Emissions Directive (2010/75/EU).

More broadly, BAT falls within the scope of the *substitution principle* that prescribes that processes, services and products should, wherever possible, be replaced with alternatives which have a lower impact on the environment. This therefore applies not only to abatement technologies but also to all kinds of products and chemicals. This design principle is therefore aligned with the idea of transmaterialisation. In many kinds of European industrial settings, and for other companies that export to Europe, the substitution principle is a driver of ongoing work to identify substances of very high concern (SVHC) in the EU chemicals regulation, REACH, and of the development of hazardous substance lists such as the SIN List (by ChemSec – the International Chemical Secretariat, an NGO).

Another concept that implicitly contains design principles for technology is *appropriate technology*. The concept was introduced (but then called *intermediate technology*) by Schumacher in his 1973 book *Small Is Beautiful*. It refers to technology that is small-scale, decentralised, labour-intensive, energy-efficient, environmentally sound, locally autonomous and people-centred. Many times, these technologies are also developed according to open-source principles and are then called open-source appropriate technology (OSAT). The concept is particularly useful for developing regions and others lacking centralised (e.g. national) infrastructure.

1.3.5 Management Principles

Another driver behind the identification of SVHCs and the SIN list is the *precautionary principle*. It is an approach to risk management that implies that precautionary measures should be taken when an activity raises threats of harm to human health or the environment, even if cause-effect relationships are not fully established scientifically.

It therefore suggests that harm should be anticipated and that a risk-averse approach is applied. In more recent interpretations, it also suggests that the burden of proof that an activity is not harmful falls on those that want to take that action. It is therefore a management principle that tells us how to deal with uncertainty. The precautionary principle is a guiding principle behind much environmental legislation today.

Other principles that are related are the *polluter pays principle* (PPP) that proposes that the party responsible for producing pollution is also responsible for paying for the damage done to the natural environment. The same thinking is partly behind the strategy of *extended producer responsibility* that requires that producers take responsibility for steps in the life cycle of a product other than those they directly own (i.e. production). This strategy has particularly influenced packaging waste legislation in many countries, e.g. members of the EU. Besides requiring that producers inform consumers regarding how to appropriately dispose of their wastes, it requires that collection and recycling systems are implemented and it therefore indirectly pushes producers to design products that are easier to recycle.

1.3.6 Critiques and Limitations of an Engineering Mindset

The principles and strategies that have been discussed in this chapter have mostly been formulated with an engineering mindset and to provide guidance to engineers. However, there are of course other ways to view the world than with an engineering perspective that are complementary and sometimes even more appropriate. This section discusses some limitations of the traditional engineering perspective that need to be kept in mind when applying the perspectives offered by industrial ecology and other principles discussed in this chapter.

Firstly, the presumptions that everything worth knowing can be observed and quantified and that the system is not greater than the sum of its parts are problematic in some contexts. Engineers often use systems analysis tools to understand various problems and find solutions to them. Systems analysis in general is often claimed to be an expression of a reductionist view of the world in the sense that it attempts to explain systems in terms of their individual, constituent parts and interactions between them. Further, it often also relies on a positivist paradigm, prevailing in natural sciences, in which knowledge is based on what can be objectively measured. A further problem is that the focus on things that can be quantified may take the focus away from aspects that cannot be quantified but that may be even more important. For example, in life cycle assessment, indicators that can be assessed quantitatively may capture only some of the environmental impacts that are relevant to include in an environmental assessment in a specific context. Climate change and energy use are impact categories that are often included in the assessment while human toxicity and ecotoxicity are rarely included, and the first two can also generally be quantified with lower levels of uncertainty than the last two. When results are interpreted, however, it is sometimes forgotten that there are other impacts that cannot be seen in the assessment results and that uncertainties may vary greatly. The few quantitative results that exist often become the sole focus of attention in the discussion on conclusions. To an uncritical eye, graphs that compare LCA results for two technologies may be interpreted as truths that are certain and conclusions may then be based on incomplete information. For technical systems and for some levels in ecological systems, the atomistic and positivist view may be useful, even necessary, for studying important phenomena. However, as soon as human social aspects are included, as is almost always the case when we are considering sustainable development, the systems we study have properties that cannot be captured by traditional engineering tools and perspectives. Then, the situation automatically gets messy as problems are ambiguous and systems have emergent properties.

Such perspectives therefore have to be complemented and sometimes even replaced by other studies. However, engineers cannot just hand these situations over to people of other professions, but rather they need to expand their toolbox and available perspectives to try to manage the situation, possibly together with other professionals and with various stakeholders that have an understanding of the situation and/or will be part of interventions aimed to improve the situation.

Engineers, therefore, cannot be trained merely to solve simple mathematical story problems, or perform highly complex or complicated modelling. They also need to get experience of working with ill-structured or wicked problems for which 'solutions' do not really exist but only ways to improve the overall situation. A traditional engineering perspective can teach us many things about the world and what we need to do, but it is not enough to 'solve' the grand challenges that we face related to e.g. climate change and poverty. One thing is clear: we will not contribute to sustainable development by only continuing business as usual, so we need to change the way different professionals, like engineers, approach contemporary problems.

1.4 Engineering and Applied Science Careers in Environment and Sustainability

Today, many engineers and scientists are professionally engaged in the management of environmental or sustainability issues in some way. Ten of the important sectors of the 'green economy' identified by Dierdorff et al. (2009) include those shown in Table 1.2.

But skills in environmental management are not only important in obviously 'green' industry sectors. Whether or not they would be a part of this list, many large companies apply some kind of environmental management system that involves people throughout the organisation in setting and following up various environmental goals, and intervening in business as usual to achieve them. Many companies also work actively within the broader scope of corporate social responsibility to e.g. make sure that people are treated in a fair and respectful way throughout the value chain. New or strengthened legislation, such as the REACH chemicals regulation in the EU or the requirement to make public sustainability reports, forces organisations to learn more about their potential impacts and act on this knowledge. Further, customers are increasingly putting pressure on producers to reduce environmental and sustainability impacts related to the products and services they buy. Therefore, environmental and sustainability concerns are today integrated into the work of engineers who design products, who work in production and in marketing and sales, or who are

Table 1.2 Key Sectors of the 'Green Economy'	
Sector	Example interventions employees work on in this sector
Agriculture and Forestry	Developing natural pesticides, implementing efficient land management of farming, forestry and aquaculture.
Energy Efficiency	Many engineers and scientists strive to increase the efficiency of buildings and factories, making energy supplies more effective, constructing 'smart grids', etc.
Environment Protection	In this sector, professionals work directly on environmental quality, including soil remediation, climate change adaptation and air quality initiatives.
Governmental and Regulatory Administration	Public and private organisations have people working in policy development and implementation for nature conservation, pollution prevention, regulation and enforcement.
Green Construction	Constructing and retrofitting residential and commercial buildings, and other major infrastructure.
Manufacturing	Developing new products and new ways to make old products can lead to reduced impacts.
Recycling and Waste Reduction	Solid waste and wastewater management, treatment and reduction, as well as processing recyclable materials.
Renewable Energy	This sector covers work on developing and using energy sources such as solar, wind, geothermal and biomass.
Research, Design, and Consulting Services	There are many 'indirect jobs' in the green economy, including employees in consulting firms and universities who have specialised in a field that enables them to service many different corporate and government organisations.
Transport	Many professionals aim to reduce the environmental impact of various modes of transportation, including trucking, cars and public transport.

managers of these engineers. To aid them, there are also many engineers who are environmental specialists and who populate various environmental and sustainability offices or are consultants to organisations that are outsourcing this specialist competence. So aside from any sense of personal engagement with sustainability issues, there are good career reasons for students of engineering and science to learn more about environmental management.

1.5 Review Questions

1. Select a sustainability-motivated project or large-scale effort (e.g. related to the shift in petrochemical industry to biobased raw materials, the introduction of circular business models in the fashion industry or the electrification of small, rural villages in Africa) and discuss which of the areas captured by the Sustainable Development Goals (SDGs) (Table 1.1) may be affected by the project, positively

or negatively. You may also look up more details on the subgoals in the SDG reports.

2. Describe the origin of the principle of circular flows in industrial ecology.

3. For one large material flow in society (e.g. soda cans, clothes, plastic bags, food) discuss how flows could become more circular. Use the suggestions in the model from the Ellen MacArthur Foundation in Figure 1.2 as inspiration.

4. For a selected environmental impact and policy context (e.g. global climate change and the global community or eutrophication in the Baltic Sea and any national government managing a coastline), suggest indicators for the whole DPSIR framework in Figure 1.3.

5. Discuss how your local household waste management could be changed to be better aligned with the waste hierarchy in Figure 1.4.

6. Describe common critiques and limitations related to systems analysis and an 'engineering mindset'.

2

The Earth System
Natural Operation and Human Impacts

Humans are just one of the many living species on the Earth. All living organisms have an impact on their immediate environment. They all need nutrients to survive and grow and they create waste that needs to be taken care of. Over time, mature ecosystems have formed, in which matter is circulated and the type of life that is contained in them is fairly constant. If the population of a species suddenly increases, it will soon be pushed back by different feedback mechanisms in the environment, for example, lack of resources, the accumulation of waste or because populations of organisms that feed on them start to grow faster. The same thing of course applies to the human population. However, human ingenuity has resulted in technological developments that have reduced our dependence on local renewable resources, and made non-renewable resources available, like various metal ores and fossil fuels. We are also less sensitive to predation nowadays. These developments offer us great opportunities compared with other species, but they also mean we are directly changing our environment. We also do it indirectly by transplanting invasive species that change ecosystems. Many of the changes we cause may be irreversible and may cause natural systems all over the world to shift into new regimes, which will change the conditions for life on Earth. Because of this large-scale, human-induced change, it has been suggested that the current part of the Holocene period should in fact be called the Anthropocene period, as human actions dominate the process of change in global ecosystems. A good understanding of the functioning of the Earth system is needed to mitigate or adapt to such changes and to know how to prevent potential future impacts. In this chapter, the Earth system is described from three different perspectives: the four environmental compartments ('spheres'), biogeochemical cycles and ecosystems.

2.1 The Four Environmental Compartments

Our environment is often said to consist of four different compartments: the atmosphere, the hydrosphere, the lithosphere and the biosphere. These compartments are not isolated from each other; they are constantly exchanging energy and materials. Also within the compartments, matter moves and may be transformed. This movement and transformation of matter can be described by biogeochemical cycles, as explained later in this chapter. Some important drivers behind such movement and transformation are (i) the second law of thermodynamics – in this case, in the absence of other drivers, natural systems degenerate towards higher entropy (disorder), (ii) the energy input from the sun (and to some extent also from e.g. the Earth's hot core and

Table 2.1 Major Constituents of the Atmosphere	
Substance	% by weight
Nitrogen	77.78
Oxygen	20.87
Argon	0.93
Carbon dioxide	0.04
Water vapour	~0.40

from gravitational interference by the moon), and (iii) life-governed processes, directed by information including but not limited to DNA, social instincts, culture and (hopefully) scientific knowledge.

2.1.1 The Atmosphere

The atmosphere contains the air that surrounds the Earth. The air consists primarily of nitrogen and oxygen and varying levels of water, some argon and carbon dioxide and trace amounts of other gases; see Table 2.1. The chemical and physical forms of these elements is of importance for the functioning of the Earth system. The atmosphere has an important role in the dispersion and transformation of pollutants emitted to air, as explained further in Chapter 3. Nitrogen in the atmosphere is primarily present as nitrogen gas (N_2), but it is also part of other compounds, including some that are critical nutrients for plant growth but that may also cause some major environmental issues. This is discussed further in Section 2.2.3 on the biogeochemical cycle of nitrogen. Atmospheric carbon dioxide (CO_2) is used in plant photosynthesis, and oxygen (oxygen gas, O_2) is a waste product of the same process. The biogeochemical cycle of carbon is further discussed in Section 2.2.2. Water exists in the atmosphere either as vapour or as droplets of varying size in clouds and fog; see the water cycle, Section 2.2.1.

The atmosphere can be divided into different layers; see Figure 2.1. Closest to the surface of the Earth and up to an altitude of about 15 km is the *troposphere*. This is the part where most of what we call 'weather' happens, and aeroplanes normally fly within the troposphere or possibly in the lower levels of the *stratosphere*, which is directly outside the *tropopause* (a 'pause' is a boundary between two atmospheric layers). The air in the troposphere is warmest at the surface of the Earth. About 30% of the incoming solar radiation (ultraviolet – UV – and visible) that penetrates the atmosphere is reflected due to the albedo of the surface or by particles and droplets in the atmosphere, and the remainder is absorbed by the Earth's surface, warming up the land and water, and is then re-emitted as infrared radiation into the atmosphere. Greenhouse gases in the atmosphere (primarily water and carbon dioxide) will absorb this radiation and re-emit it in all directions; see Figure 2.2. This 'natural greenhouse effect' slows down heat loss from the Earth system and increases the air temperature. This makes the Earth's surface about 30–35°C warmer than it would be without this effect and gives us a comfortable average surface air temperature of about 15°C. As

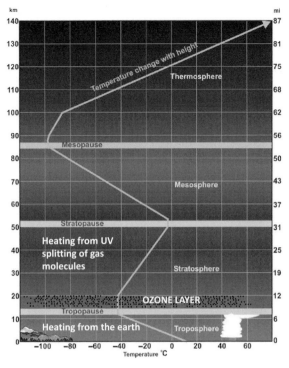

Figure 2.1   Layers of the atmosphere. Based on a figure from the National Weather Service, n.d.

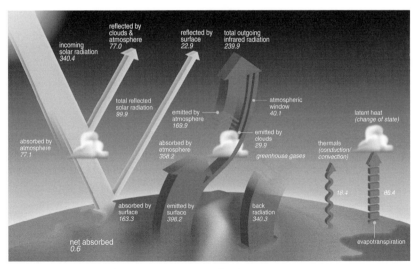

Figure 2.2   The global radiation balance. Values are fluxes in W/m². Based on a figure from NASA, 2017

there is a balance between incoming solar radiation and heat radiation that leaves the atmosphere, we have a fairly constant average temperature at the surface of the Earth under natural conditions (see further information on climate change in Chapter 3).

The air within the troposphere is continuously mixed under normal conditions on account of winds and convection currents. Convection occurs because air expands when hot, reducing its density. So it rises above 'heavier' (denser) air that blows in to take its place. However, there are temporary, regional situations when a *temperature inversion* may arise. If the land surface receives very little radiation from the sun (e.g. overnight and/or during winter), a layer of colder air, perhaps a few hundred metres thick, may be trapped under a warmer layer. Inversions like this may lead to problems when large emission sources are located within such areas, as emissions are not rapidly diluted; see further details in Section 3.4.

The stable layering of the atmosphere into the troposphere and the stratosphere (at 15–40 km) is also a result of such a temperature inversion; see the temperature gradient in Figure 2.1. The warming that takes place in the upper levels of the stratosphere is because of the strong solar radiation that splits nitrogen and oxygen gas molecules into free radicals. This splitting absorbs certain bandwidths of UV radiation, protecting the troposphere and us from that radiation and converting the energy to heat. The further down in the stratosphere we look, the more of this radiation has been stopped and the warming effect is lower. In the lower stratosphere, a low but stable concentration of ozone (O_3) forms as a result of the splitting of oxygen gas (O_2) molecules into free oxygen atoms that can combine with oxygen gas molecules to form ozone. The normal concentration of ozone in the stratosphere is an equilibrium concentration that is determined by several different reactions, see further details in Section 2.1.1.

We call this part of the stratosphere that contains small amounts of ozone the 'ozone layer'. The temperature inversion reduces mixing of air within the stratosphere and, in particular, creates a barrier to the exchange of gases between the troposphere and the stratosphere.

The temperature of the gases in the thermosphere can exceed 2000°C on account of the absorption of UV and X-ray radiation above an altitude of about 85 km. However, an astronaut in this region would not feel hot, as the air pressure is so extremely low (about 1 Pa at 85 km) that heat transfer is limited.

The rotation of the Earth, inertial forces in air masses and the effects of rising air currents at the equator and elsewhere create the average wind patterns around the Earth shown in Figure 2.3. These wind patterns transfer water and various pollutants between regions, and they are a key factor behind both the so-called ozone hole (see Section 3.5) and behind the so-called grasshopper effect (also called *global distillation*; see Section 3.7).

2.1.2 The Hydrosphere

Most of the surface of the Earth, 71%, is covered by water, primarily as salty water in seas and oceans (97.4%). Only 2.6% of the water around us is freshwater, most of which is in ice and groundwater. Less than 0.007% of all freshwater is in rivers, lakes and in the atmosphere, and only 0.0001% is contained in biological organisms and manufactured products.

Oceanic circulation has two primary causes. Rapid surface currents are primarily driven by wind. Although thermohaline circulation, driven by temperature and salt

different organisms. It also holds much of the water that the plants need, which would otherwise drain away.

2.1.4 The Biosphere

The biosphere encompasses life on Earth in its various life forms. The different organisms take part in the cycling of matter within the Earth system in biogeochemical cycles (see Section 2.2) and can be said to be organised into ecosystems (Section 2.3). The organisms have a large influence on the composition of the Earth's atmosphere and hydrosphere.

Biologists tell us that the first single-celled organisms formed about 3–4 billion years ago and that all other organisms derived from these first life forms through evolution. It is not known how many species exist, but it is believed that the total number could be about 1 trillion. About 1.6 million species have been listed in databases. It is estimated that more than 99% of all species that ever lived on Earth are now extinct. Both extinction and speciation have occurred since life on Earth began. Since the rapid proliferation of multicellular organisms about 541 million years ago, which we call the 'Cambrian explosion', there have been certain periods of mass extinction caused by factors including volcanism and asteroid impacts. The current impact of human society on biodiversity has been called the 'sixth mass extinction' since that time.

It is estimated that the total mass of life on Earth is in the range of several billions of tonnes of dry matter, of which the vast majority is made up of plants. However, in recent years, the total mass of underground microorganisms has been estimated to make up as much as 30% of our planet's biomass.

Different species have adapted to fill certain ecological niches. Some specialised life forms exist even in very harsh places of the Earth, such as in hydrothermal vents in the deep ocean where some microbes have been proven to thrive at highly elevated pressures and temperatures. One example is *Methanopyrus kandleri*, a hyperthermophile that can survive and reproduce at 122°C. There are even examples (e.g. *Kineococcus radiotolerans*) that can live in nuclear waste dumps and survive radiation doses that would bring certain death to humans.

About 99% of a living cell consists of just a few elements: hydrogen, oxygen, nitrogen, carbon, sulphur and phosphorus. About 70% of the wet weight of a cell is water. The dry weight is made up primarily of organic macromolecules, and these are unique to the living organisms. Most of the organic macromolecules in a cell belong to one of four major classes: nucleic acids, carbohydrates, proteins and lipids. About 1% of the cell mass is made up of inorganic ions, including sodium (Na^+), potassium (K^+), magnesium (Mg^{2+}), calcium (Ca^{2+}), phosphate (HPO_4^{2-}), chloride (Cl^-) and bicarbonate (HCO_3^-).

In 1934, American oceanographer Alfred Redfield revealed what we now call the 'Redfield Ratio'. He claimed that the average ratio of carbon to nitrogen to phosphorus in marine phytoplankton is 106:16:1. This is useful because given data about the concentration of these elements in different ecosystems, we can use this ratio to identify which element is locally limiting growth. Typically nitrogen is the limiting element in oceans while terrestrial and freshwater ecosystems are limited by

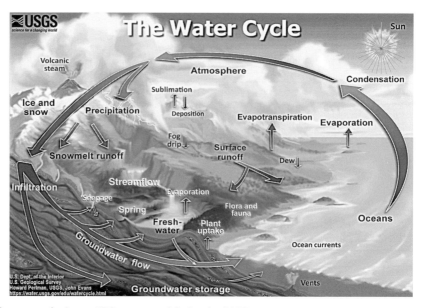

Figure 2.6   The water cycle. Source: US Department of the Interior (2017)

phosphorus, but it can also be none of those three elements. Scientists have subsequently added other elements to the end of the ratio (e.g. iron: 106:16:1:0.001) as a way to express this. It has been proposed that adding iron to the ocean is a way to stimulate growth and thereby pull the other elements out of the atmosphere.

2.2 Biogeochemical Cycles

The movement of matter in the Earth system is often represented by biogeochemical cycles. In fact, such cycles can be identified for any substance. The most commonly described biogeochemical cycles are water, carbon and nitrogen. In this book we prioritise description of these cycles because they are the ones that are the most essential to life, their flows are very large and they can be disturbed in important ways by human interaction.

2.2.1 The Water Cycle

The water cycle or the hydrological cycle, Figure 2.6, is primarily driven by energy from the sun that evaporates water from the surface of the Earth. This water rises into the atmosphere, cools down, condenses into rain or snow and falls again to the surface as precipitation. The water cycle constantly renews our access to freshwater, and it also makes sure that our hydropower reservoirs are refilled. Glaciers may hold water over long periods of time and they have an important role in the seasonal regulation of freshwater access downstream. Furthermore, the cycling of water in and out of the

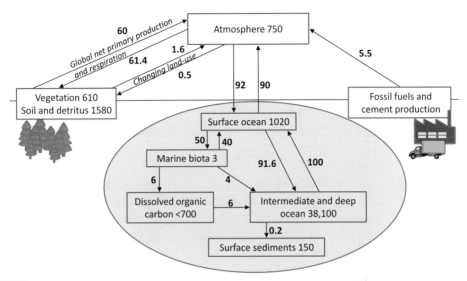

Figure 2.7 The carbon cycle, with global stocks and flows. Stocks (boxes) in GtC and fluxes (arrows) in GtC/yr. Stocks of fossil fuels etc. in the lithosphere are not shown due to large uncertainties

atmosphere is a significant aspect of the weather patterns and water vapour is also a key greenhouse gas.

The water cycle has obvious effects on the biosphere. Less well understood and appreciated are the influences the biosphere has on the water cycle. For example, it has been hypothesised that marine phytoplankton generate dimethylsulfide gas, which helps produce sulphate aerosols that act as cloud condensation nuclei. The clouds blow on shore, shading the land (reducing terrestrial evaporation) and bringing further precipitation – in this way, onshore air temperatures and precipitation patterns are influenced by organisms living in the sea!

2.2.2 The Carbon Cycle

The carbon cycle encompasses two major exchanges between environmental compartments, and several others (Figure 2.7). The first is the exchange between hydrosphere and atmosphere in which carbon dioxide in the atmosphere equilibrates with carbon dioxide dissolved in water, which in turn is in equilibrium with other reactions in the water involving bicarbonate (HCO_3^-) and carbonic acid (H_2CO_3). Some examples are:

$$CO_2(g) + H_2O \rightleftharpoons CO_2(aq) + H_2O \qquad\qquad \text{dissolution} \qquad (2.1)$$

$$CO_2(aq) + H_2O \rightleftharpoons H_2CO_3(aq) \rightleftharpoons H^+ + HCO_3^-(aq) \rightleftharpoons 2H^+ + CO_3^{2-}(aq) \text{ deprotonation} \qquad (2.2)$$

$$Ca^{2+}(aq) + CO_3^{2-}(aq) \rightleftharpoons CaCO_3(s) \qquad \text{precipitation} \qquad (2.3)$$

The lower the temperature, the more carbon dioxide can be dissolved. The other major exchange process involves carbon dioxide in the atmosphere and carbon in the biosphere, where photosynthesis and respiration are the reactions of interest. In the photosynthesis in green plants, carbon dioxide and water react to yield glucose and oxygen, and through the respiration process in all organisms, glucose and oxygen react to yield carbon dioxide and water.

$$6\,CO_2 + 6\,H_2O + light \rightarrow C_6H_{12}O_6 + 6\,O_2 \quad \text{photosynthesis} \qquad (2.4)$$

$$C_6H_{12}O_6 + 6\,O_2 \rightarrow 6\,CO_2 + 6\,H_2O + energy \quad \text{respiration} \qquad (2.5)$$

In this way, carbon is constantly cycled between biomass and atmosphere. It is tied up as biomass grows and it is released as it dies. For this reason, global carbon dioxide concentrations are a little higher during wintertime in the northern hemisphere because of the release of carbon from rotting leaves in the large deciduous forests in Eurasia and North America.

Large amounts of carbon are tied up in fossil fuel reserves in the lithosphere. These do not naturally participate in the rapid cycling of carbon, so the engineered release of carbon from these into the atmosphere can therefore increase the normal background concentrations in the atmosphere. An understanding of the carbon cycle is important for understanding climate change and for finding effective mitigation measures; see Section 3.6. Further, it is central in understanding ocean acidification; see Section 3.2.

2.2.3 The Nitrogen Cycle

The biosphere (and in particular microbial life) plays a critical role in the operation of the natural nitrogen cycle (Figure 2.8). Bacteria and also certain plants that are in symbiosis with bacteria (e.g. legumes) are responsible for the 'fixation' of nitrogen gas from the atmosphere (converting N_2 into ammonia, NH_3). Microbes living in the soil and oceans are also important for their role in oxidising ammonium ions (NH_4^+) emitted from decaying organic matter to nitrite (NO_2^-) and then to nitrate (NO_3^-). This sequence of reactions is called 'nitrification'. Living organisms typically absorb nitrate and ammonium, add the nitrogen in them into molecules like amino acids, nucleic acids and chlorophyll and incorporate them into organic matter. The microbial process of 'denitrification' converts nitrate back to nitrogen gas. Denitrification is a respiratory strategy in environments lacking oxygen gas as a terminal electron acceptor (see the description of energy conversion in living things in Section 5.2.1 to put this in context). Nitrous oxide (N_2O – an important greenhouse gas) can be produced via incomplete nitrification or as an intermediate product during denitrification.

Nitrogen is the growth-limiting plant nutrient in many environments. Industrial nitrogen fixation (the Haber-Bosch process) is used to produce fertilisers that increase agricultural production. The other side of the coin is that excess nitrogen can create major environmental problems on different scales depending on the stability and mobility of the nitrogen-containing molecules involved. The many different forms of nitrogen that exist in water (for example nitrate) and in air

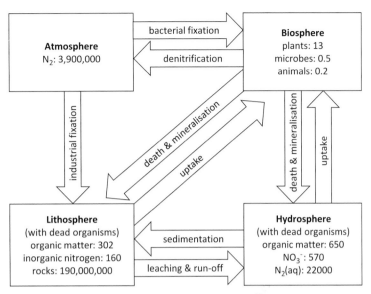

Figure 2.8 The nitrogen cycle with major flows and stocks indicated (units = 10^{15} g nitrogen)

(for example ammonia and nitrous oxide), and that are all part of the nitrogen cycle, have important roles in several different environmental problems, as is further described in Chapter 3.

2.3 Ecosystems

This chapter previously discussed the four 'spheres' and the biogeochemical cycles as two different ways to describe the environment. A third possibility is to describe the environment in terms of ecosystems. Instead of focussing on different compartments or the movement of certain elements or substances, the ecosystem perspective instead takes its starting point in how life has organised itself in different environments.

2.3.1 Organism Functional Types

The organisms in an ecosystem can be sorted into three generic types depending on their ecosystem function: producers, consumers and decomposers; see Figure 2.9. These organism types together form food chains, or food webs.

The producers, mostly green plants, utilise energy from the surroundings, most often in the form of solar energy via photosynthesis, to create energy-rich molecules that fuel the ecosystem. The other organism types in the ecosystem rely on them for their survival. The producers are food for primary consumers (called herbivores, e.g. caterpillars, cows), which may be prey for secondary consumers (called carnivores, e.g. falcons, lions), and some consumers may feed on both groups (and are then called omnivores, e.g. bears, humans). The decomposers (detritivores, e.g. fungi, woodlice) utilise the remaining energy in dead organisms or waste from producers or consumers and simultaneously

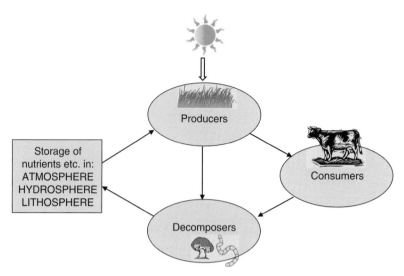

Figure 2.9   The functional organism types in an ecosystem

return the matter to soil, water and air. In this way, a circular flow of nutrients is generated.

2.3.2 Characteristics of Food Chains

Ecosystems or food chains can also be illustrated in terms of how numbers of individuals, or quantities of biomass and energy are distributed between different trophic levels. This also varies between different ecosystems. The energy pyramid is always a true pyramid; see Figure 2.10. Anything else would be impossible as energy is always lost between different levels. The producers (grass or phytoplankton in the example in the figure) are the components that introduce chemical energy into the ecosystem, and they are therefore at the bottom (primary production). The primary consumers (rabbit or zooplankton) feed on the producers, the secondary consumers (herring or fox) on the primary consumers and so on. The energy loss is typically about 90% between trophic levels in the food pyramid, i.e. only 10% of what is consumed effectively generates new biomass at the higher trophic level. This is one of the major arguments behind vegetarianism: if we consume from the lowest trophic level, we can feed more individuals. Of course, many consumer organisms (cats, for instance) do not have that choice (because cats can get some essential nutrients, like the amino acid taurine, only from meat).

However, looking instead at the total biomass or the number of individuals, the relations will depend on many different things such as the size, life length and reproduction rates of organisms (e.g. compare trees to phytoplankton populations).

The size of populations is regulated in different ways. Bottom-up control means that access to resources (nutrients, space, etc.) regulates the population size while top-down control means that the relationship between predators and their prey regulates population sizes.

The reproductive patterns of a species are of course also very important. Species have historically been divided into r-strategists and K-strategists, depending on

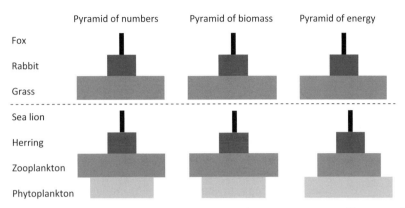

Figure 2.10 Food chain pyramids, illustrating typical relation for numbers of organisms, quantities of biomass and energy in terrestrial (upper pyramids) vs aquatic (lower pyramids) ecosystems

whether they produce a high quantity of offspring but offer low-quality parental care or vice versa. The pronumerals are taken from the simplified Verhulst model of population dynamics:

$$\frac{dN}{dt} = rN\left(1 - \frac{N}{K}\right) \tag{2.6}$$

where N is the population of the species, t is time, r is a rate coefficient and K is the carrying capacity of the ecosystem. While less popular than they were in the 1970s, the r and K labels are still used to describe how the life stories of organisms differ from each other. R-strategists or 'r-selected species' reproduce quickly by having short gestation periods and large numbers of offspring per gestation. They are generally small (bacteria, insects and mice). By manifesting a high r in the Verhulst model they take advantage of changing environments, but this can also result in overpopulation, especially when environmental conditions change for the worse. On the other hand, 'K-selected species' generally produce fewer offspring per reproductive cycle and the cycles are longer, but this makes sense when N is close to K, and the offspring can often be expected to survive to their own sexual maturity. Having time to grow, K-strategists are also typically larger (whales, humans and elephants).

Whether an organism thrives in an ecosystem depends in part on abiotic (non-living) factors. Different abiotic components of an ecosystem matter to different species, e.g. depending on whether it is a terrestrial or an aquatic species that is considered. In terrestrial ecosystems, important abiotic factors may include things like sunlight, temperature, wind, rain, altitude, type of soil, pH, nutrient availability, etc. In an aquatic ecosystem, they may include things like water depth, turbidity, sunlight, temperature, salinity, dissolved oxygen, etc. These factors may exhibit regular variation (e.g. daily, lunar, annual) or apparently random dynamics, or a combination of both.

2.3.3 Ecosystem Resilience and Biodiversity

When a disturbance occurs in an ecosystem, the term *resilience* is used to describe its capacity to resist damage and recover, or adapt so that functionality is restored. Such disturbances can be natural, e.g. droughts, storms and fires, and they can be caused by human activities, such as deforestation or the introduction of foreign species. The resilience of an ecosystem depends e.g. on the organism types and their relations. If some organisms are highly specialised and there are very strong interdependencies between certain organism types, that ecosystem is more vulnerable and cannot recover from or adapt to change as easily as a system where organisms can survive temporarily or permanently by e.g. selecting other food sources. The tolerance of different organisms to changes in abiotic factors also varies greatly, and their tolerance range for one factor may become narrower if they are suffering from stress due to a disturbance in another factor. For example, exposure to sulphur dioxide pollution disturbs the function of the stomata (gas exchange pores) of leaves, leaving the plant more sensitive to droughts because of larger transpiration losses and also more sensitive to the intrusion of damaging tropospheric ozone.

On an ecosystem level, biodiversity is a very important factor behind resilience. Biodiversity is normally defined on at least two levels (genetic and functional) but often up to four different levels:

• Genetic diversity: This is the genetic diversity within and between populations of a species in an ecosystem. When this is low, there is a larger risk that all individuals of that species will be lost when a severe disturbance occurs. This may lead to species extinction, at least locally, and if that species has a key role in the ecosystem functioning, this can be critical.
• Species diversity: Different species have different roles and some may compete for similar niches, but this also gives redundancy in case something should happen to one of the species. Therefore, species diversity focusses on the number of different species.
• Functional diversity: Functional diversity instead focusses on the range of functions that the species provide. If all the species in an ecosystem do the same things, fewer ecosystem services are delivered.
• Landscape diversity: This relates to the types of biomes that exist in the world. The existence of many different types of ecosystems reduces the vulnerability to damage to one type of biome on a global level.

When conditions change beyond certain thresholds, an ecosystem may be pushed into a regime shift. One example is how a rainforest may shift into a savannah after logging because the moisture recycling feedback is weakened. Some regime shifts are practically irreversible.

2.4 Human Dependence on Ecosystems

We may forget it sometimes, but, ecologically, we are just one of the consumer species in the global ecosystem, and we have to rely completely on functioning ecosystems and on the abiotic resources that can be extracted from the natural world. We sometimes use the term *natural capital* for the various natural resources and services in our environment that can potentially be exploited to create value to us.

Table 2.4 Taxonomy for Different Types of Natural Resources				
Taxonomy I	**Taxonomy II**	**Examples**	**Potential issues**	**Management strategies**
Non-renewable	Stock-type	Fossil fuels, minerals	Scarcity	Recycling
Renewable	Fund-type	Wood, fish	Extraction rates; functionality of renewal mechanisms	Regulation of resource extraction; protection of renewal mechanisms
	Flowing; flow-type	Sunlight, wind, rain	Local competition; life cycle impact of extraction	Regulation of resource extraction; more efficient extraction

Natural resources, in turn, are often divided into renewable and non-renewable resources, depending on whether there are mechanisms that will renew them so that their use will not automatically lead to depletion. The fossil fuels and the various minerals are examples of non-renewable resources while wood in a forest and water in a river are renewable. For the non-renewable resources, their general scarcity, their accessibility and quality, the available extraction technology and the degree to which we organise for circular flows (recycling) will determine the rate of depletion and therefore also how critical extraction of the resource is. Renewable resources differ in terms of whether the renewal mechanism itself can be disturbed and therefore, a more useful taxonomy for all resources may be the division into stock-, fund- and flow-type resources; see Table 2.4. In this taxonomy, the stock-type resources are the non-renewable ones – there is a certain stock that will not be renewed once it has been extracted. In the idea of fund-type resources, there is a renewal mechanism analogous to the idea of a financial fund that pays annual dividends. If the fund is well managed and extraction rates are not too high, these resources will be renewed, and potentially at even higher rates than is natural if best practice management is implemented. Forests, for example, are more productive at a certain extraction rate than above or below it. The flow-type resources have renewal mechanisms that in principle cannot be disturbed. Solar energy will always enter the Earth system (on human timescales) and set water and air masses into motion.

The natural *services*, however, are not always as tangible to us even if they are at least as important as the natural *resources*. In recent years, many have found the term *ecosystem services* useful to describe the services that ecosystems provide us, beyond agricultural productivity, timber, fish and other resources that are commodities on the market. In fact, this term encompasses at least both natural services and fund-type natural resources. The most well-known description of ecosystem services is the one that was used in the Millennium Ecosystem Assessment Report in 2005; see Figure 2.11. What is also very interesting about this description is that it connects the ecosystem services to different constituents of human well-being. Remember that both the concept of ecosystem services and the concept of well-being have here been interpreted and described in a certain way, but both these concepts are constantly

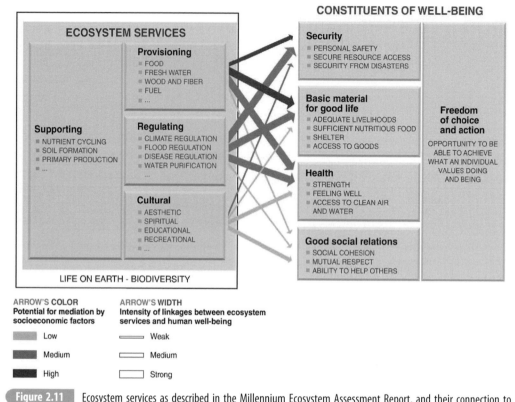

CONSTITUENTS OF WELL-BEING

ECOSYSTEM SERVICES

Provisioning
- FOOD
- FRESH WATER
- WOOD AND FIBER
- FUEL
- ...

Supporting
- NUTRIENT CYCLING
- SOIL FORMATION
- PRIMARY PRODUCTION
- ...

Regulating
- CLIMATE REGULATION
- FLOOD REGULATION
- DISEASE REGULATION
- WATER PURIFICATION
- ...

Cultural
- AESTHETIC
- SPIRITUAL
- EDUCATIONAL
- RECREATIONAL
- ...

LIFE ON EARTH - BIODIVERSITY

Security
- PERSONAL SAFETY
- SECURE RESOURCE ACCESS
- SECURITY FROM DISASTERS

Basic material for good life
- ADEQUATE LIVELIHOODS
- SUFFICIENT NUTRITIOUS FOOD
- SHELTER
- ACCESS TO GOODS

Health
- STRENGTH
- FEELING WELL
- ACCESS TO CLEAN AIR AND WATER

Good social relations
- SOCIAL COHESION
- MUTUAL RESPECT
- ABILITY TO HELP OTHERS

Freedom of choice and action

OPPORTUNITY TO BE ABLE TO ACHIEVE WHAT AN INDIVIDUAL VALUES DOING AND BEING

ARROW'S COLOR
Potential for mediation by socioeconomic factors
- Low
- Medium
- High

ARROW'S WIDTH
Intensity of linkages between ecosystem services and human well-being
- Weak
- Medium
- Strong

Figure 2.11 Ecosystem services as described in the Millennium Ecosystem Assessment Report, and their connection to constituents of human well-being. Source: Millennium Ecosystem Assessment (2005)

reinterpreted by different groups and the models that are presented can be quite different.

As shown in Figure 2.11, ecosystem services are commonly divided into four groups. *Provisioning services* are those that directly provide people food, water, fuels and materials of different kinds. This is where we find the fund-type natural resources. *Regulating services* are those that provide homeostasis (they regulate towards a stable state) for our living conditions, for example, regulating the climate, the river flows and the quality of air, water and soil, or pollinate crops. *Supporting services* include those responsible for primary production and providing living spaces for plants or animals. Finally, there are *cultural services* that are non-material and that we make use of in teaching and learning about the environment, in ecotourism, in outdoor sports and for aesthetic and spiritual purposes. Figure 2.11 also shows how strong the linkage is to different constituents of human well-being, and the potential for mediation by socioeconomic factors, for example taxes. It is clear that it is the provisioning and regulating services that have the strongest connection to human well-being and also the strongest potential for mediation, and the constituents of well-being that are affected the most are security, basic material for good life and health.

Recent and ongoing efforts that have made attempts to value ecosystem services in monetary terms include the global Economics of Ecosystems and Biodiversity initiative that is focussed on 'making nature's values visible' (www.teebweb.org).

A related initiative is the United Nation's System of Environmental-Economic Accounting (https://unstats.un.org/unsd/envaccounting/eea_project/default.asp). Efforts to monetise ecosystem services and environmental damage have been criticised because they risk excluding aspects for which an economic value cannot be established, while increasing focus on what can be included. The more challenging aspects may then fall even deeper into political oblivion, allowing the ecosystems to be more severely exploited and downgrading their capacity to deliver services.

2.5 Human Impacts on Ecosystems

All species affect their surroundings, and humans are no exception. However, human impacts have increased considerably over time as human society has developed and the global population has increased. Hunter-gatherers and simple agricultural societies that preceded industrial society (and are still present) subsisted primarily on the resources that the local environment could produce. Today, our globalised, high-technology society allows for trade and transportation all over the world, and we have learnt to use resources in the Earth's crust that have not been accessible to earlier generations and other organisms. In this way, we have made ourselves independent of energy and material provisioning by local ecosystems. But not only have these powers that we now possess been used to generate wealth and well-being, there are also increasing signs of environmental disturbance and resource depletion.

The current state of ecosystems and the role humans play in ecosystem change is difficult to assess because ecosystems are very complex and they exhibit natural variability, but there have been some attempts that have gained much attention. Some of these are summarised in the next section. This chapter ends with a section that describes some disturbances of ecosystems while the next chapter in the book is fully dedicated to impacts of chemical pollution.

2.5.1 The State of Ecosystems in the World

The Millennium Ecosystem Assessment was initiated by United Nations Secretary-General Kofi Annan and was performed between 2001 and 2005. The assessment involved the work of more than 1,360 experts worldwide (www.millenniumassess ment.org/en/About.html). The assessment was a 'state-of-the-art scientific appraisal of the condition and trends in the world's ecosystems and the services they provide, as well as the scientific basis for action to conserve and use them sustainably'. The main findings were (Millennium Ecosystem Assessment, 2005):

- Over the past 50 years, humans have changed ecosystems more rapidly and extensively than in any comparable period of time in human history, largely to meet rapidly growing demands for food, fresh water, timber, fiber and fuel. This has resulted in a substantial and largely irreversible loss in the diversity of life on Earth.
- The changes that have been made to ecosystems have contributed to substantial net gains in human well-being and economic development, but these gains have been

achieved at growing costs in the form of the degradation of many ecosystem services, increased risks of nonlinear changes, and the exacerbation of poverty for some groups of people. These problems, unless addressed, will substantially diminish the benefits that future generations obtain from ecosystems.

- The degradation of ecosystem services could grow significantly worse during the first half of this century and is a barrier to achieving the Millennium Development Goals [NB: these have now been succeeded by the Sustainable Development Goals].
- The challenge of reversing the degradation of ecosystems while meeting increasing demands for services can be partially met under some scenarios considered by the MA [Millennium Ecosystem Assessment], but will involve significant changes in policies, institutions and practices that are not currently under way. Many options exist to conserve or enhance specific ecosystem services in ways that reduce negative trade-offs or that provide positive synergies with other ecosystem services.

Although the assessment showed a quite disturbing situation in which humans are depleting natural capital, it also showed that it is not yet too late to reverse this trend, but that the needed changes in policy and practice are substantial and were not under way when the assessment was done.

The Intergovernmental Panel on Climate Change (IPCC) was set up in 1988 by the World Meteorological Organization (WMO) and the United Nations Environment Programme (UNEP) to 'provide policymakers with regular assessments of the scientific basis of climate change, its impacts and future risks, and options for adaptation and mitigation'. Every five years, it publishes an *assessment report*, written by hundreds of leading scientists who volunteer their time and expertise to review what has been published in scientific literature. The fifth assessment report was finalised in November 2014 (e.g. IPCC, 2014). The main conclusions were the following (www.ipcc.ch/pdf/assessment-report/ar5/syr/AR5_SYR_FINAL_SPM.pdf):

- Human influence on the climate system is clear, and recent anthropogenic emissions of greenhouse gases are the highest in history. Recent climate changes have had widespread impacts on human and natural systems.
- Continued emission of greenhouse gases will cause further warming and long-lasting changes in all components of the climate system, increasing the likelihood of severe, pervasive and irreversible impacts for people and ecosystems. Limiting climate change would require substantial and sustained reductions in greenhouse gas emissions, which, together with adaptation, can limit climate change risks.
- Adaptation and mitigation are complementary strategies for reducing and managing the risks of climate change. Substantial emissions reductions over the next few decades can reduce climate risks in the 21st century and beyond, increase prospects for effective adaptation, reduce the costs and challenges of mitigation in the longer term and contribute to climate-resilient pathways for sustainable development.
- Many adaptation and mitigation options can help address climate change, but no single option is sufficient by itself. Effective implementation depends on policies and cooperation at all scales and can be enhanced through integrated responses that link adaptation and mitigation with other societal objectives.

Like the Millennium Ecosystem Assessment, this report shows how humans are fundamentally changing the natural systems and that it is urgent to address these impacts by adaptation and mitigation. While there has been resistance to this message from some quarters, sometimes referred to as 'climate change denial', it is also worth pointing out here that seven journal papers from 2004 to 2015 assessed the scientific consensus on anthropogenic global warming, and showed a very high level of agreement (in chronological order): 100%, 97%, 97%, 97%, 91%, 93%, 97% in support of the consensus (Cook et al., 2016).

Since 1998, the World Wide Fund for Nature (an NGO) has published the biannual *Living Planet Report*. This is done in collaboration with the Global Footprint Network (an independent think tank) and the Zoological Society of London (ZSL) (a charity organisation). The report contains assessments of the so-called ecological footprint (the direct and indirect area use of human activities) and the Living Planet Index (biodiversity impacts). The major conclusions reported in 2016 (WWF, 2016) were (http://awsassets.panda.org/downloads/lpr_living_planet_report_2016_summary.pdf):

• The size and scale of the human enterprise have grown exponentially since the mid-20th century. As a result, nature and the services it provides to humanity are subject to increasing risk. Scientists suggest that we have transitioned from the Holocene into a new geological epoch calling it the 'Anthropocene'.
• The future of many living organisms is now in question. Species populations of vertebrate animals have decreased in abundance by 58 per cent between 1970 and 2012. The most common threat to declining animal populations is the loss and degradation of habitat.
• Increasingly, people are victims of the deteriorating state of nature: Without action the Earth will become much less hospitable to our modern globalized society. Humans have already pushed four planetary systems beyond the safe limit of their safe operating space.
• By 2012, the biocapacity equivalent of 1.6 Earths was needed to provide the natural resources and services humanity consumed in that year.
• To maintain nature in all of its many forms and functions and to create an equitable home for people on a finite planet, a basic understanding must inform development strategies, economic models, business models and lifestyle choices. We have only one planet and its natural capital is limited. A shared link between humanity and nature could include a profound change that will allow all life to thrive in the Anthropocene.

The 'planetary systems' mentioned here refer to other interesting work by a group of scientists within Earth and environmental sciences, led by Johan Rockström from the Stockholm Resilience Centre and Will Steffen from the Australian National University. In 2009, they proposed nine *planetary boundaries* as a central concept in an Earth system framework. The researchers behind this framework focussed on what they think are important aspects of ecosystem functioning or 'planetary life support systems' essential for human survival. They identified for each of them, as far as possible, control variables with boundaries that represent risk of 'irreversible and abrupt environmental change'. The boundaries can help define a 'safe space for human development'.

The group has published several assessments of the state of Earth systems in relation to these boundaries. In a journal paper from 2015, Steffen et al. provided the assessment reported in Table 2.5. Their estimates indicated that for five assessed areas, representing four Earth systems – climate change, biosphere integrity, land-system change and biogeochemical flows – safe boundaries had been transgressed so that the functioning of the human life support system is now at increasing or high risk. This, in turn, points to land use in agriculture and our use of fossil fuels as activities with very important impacts on the natural systems. The researchers emphasised that the boundaries are 'rough, first estimates only, surrounded by large uncertainties and knowledge gaps', and that they interact in ways that are complex and not well understood. They also pointed out that some of the boundaries were not yet under-stood well enough for an assessment of the potential transgression as indicated in Table 2.5. Novel entities include chemicals, nanoparticles and genetically modified organisms (GMOs), all of which have recently been introduced in the Earth system.

We can conclude that in our efforts to fulfil human needs, which are not yet fully successful at the global level, we exploit nature in various ways. The mechanisms by which this exploitation happens and the types of impacts that it creates are manifold. Many of the mechanisms and impacts that have been highlighted by the IPCC and Steffen et al. are described in terms of chemical processes (for example, the emission of greenhouse gases to the atmosphere; the eutrophication of aquatic ecosystems by nutrients). This book recognises the value of a chemical process perspective of sustainability by emphasising system analytical tools (Chapter 7) that have their roots in chemical engineering, and biogeochemical approaches to understanding environmental problems (Chapters 3 and 4). There are other perspectives on envir-onmental sustainability, and so the final part of this chapter describes some other types of environmental impacts that result from human activities. When we use the term *environmental impact*, we may mean a number of different things, and this is heavily affected by whether we see the environment primarily as a life support system for mankind or whether we think that changes in the environment are important to avoid anyway, for the sake of other species or something else. We may also include resource depletion in this term, as well as direct impacts on human health. In what follows, we discuss some impacts on ecosystem functioning and services that are the results of other stressors than chemical pollution. Chemical stressors will instead be discussed in detail in Chapters 3 and 4.

2.5.2 Non-Chemical Stressors on Ecosystems

As earlier discussed, the functioning of ecosystems depends on different abiotic factors and on certain conditions within the living parts of ecosystems. Humans therefore have the ability to disturb ecosystem functioning in many different ways. The way we use land and water resources will have a large impact on the functioning of ecosystems. From the point of view of their ability to cause species extinctions, some of the key ecosystem stressors are: direct habitat change (e.g. deforestation), overexploitation by humans and the introduction of invasive species.

When we decide to turn natural ecosystems into cultivated ecosystems, we often greatly change the conditions, which has a large effect on e.g. biodiversity.

Table 2.5 The Planetary Boundaries and the Assessment Provided in Steffen et al. (2015)

Earth systems		Below boundary (safe)	In zone of uncertainty (increasing risk)	Beyond zone of uncertainty (high risk)
Climate change			x	
Novel entities		Boundary not yet quantified		
Stratospheric ozone depletion		x		
Atmospheric aerosol loading		Boundary not yet quantified		
Ocean acidification		x		
Biogeochemical flows	Nitrogen			x
	Phosphorus			x
Freshwater use		x		
Land-system change			x	
Biosphere integrity	Functional diversity	Boundary not yet quantified		
	Genetic diversity			x

Since industrial agricultural systems often cultivate only one crop, such as wheat, rape seed, maize, cotton, oil palm, spruce or eucalypt, we sometimes talk about such cultivated ecosystems as being monocultures. Having one primary crop will unfortunately increase the risk of soil exhaustion and it will increase the sensitivity to attacks from pests of various kind, pests that are specialised to the crop in focus. A similar risk exists when intensive agricultural practices involve keeping animals at high densities – apart from the potential cruelty of these practices, they can necessitate excessive use of medicines to prevent the outbreak of microbial diseases among the animals. To reduce the risk of pests, agricultural and forestry practices can rotate between different crops or a combination of crops in the same area. Regardless, agriculture leads to habitat loss for many species that occupied the area before the land-use change. Therefore, we normally set aside some areas as nature reserves. It is estimated that about 15% of global land area is protected in this way, while for marine environments, only a little more than 1% of their global area is protected.

Habitat loss results not only from the change into other productive ecosystems but also from our increasing use of land for cities, roads, landfills and industries, and because of desertification, soil salinisation and soil erosion. Desertification is often a result of overexploitation of soil in relatively dry areas, and often in combination with climate change. When nutrients and organic matter are depleted, vegetation cover is lost, and the soil that remains is soon eroded by wind and water. It can also be the result of salinisation of soil after extensive irrigation in dry areas that have brackish or salty groundwater. When groundwater tables are raised due to irrigation and water evaporates in the upper layers, it leaves increased concentrations of salt in the soil that may harm the vegetation and thereby start a desertification process.

Sensitive monocultures, soil exhaustion from intense use and habitat loss are only some of the problems that may arise from our use of land. Soil compaction is another problem and it is caused by heavy machines used in agriculture or forestry. When the soil is compacted, the pores in the soil are closed, leading to increased water run-off and soil erosion. Another example of a problem is how excessive freshwater with-drawal from coastal aquifers may lead to seawater intrusion, changing the conditions for groundwater-dependent ecosystems.

Furthermore, we can affect the populations of species in a more direct way. When we forage, harvest or fish, we must be careful not to extract too large a part of the population so that the reproductive abilities for the species are jeopardised. This is particularly challenging for K-strategists, and it is for example the reason why there is a maximum size limit for lobster fishing in some parts of the world (to protect breeding stock). Fishing quotas are designed with the purpose of allowing for healthy regrowth of the stock. On the other hand, some populations need to be kept at a low enough level so as to not cause problems. Humans may then have an important role to play as top predators (by hunting) that keep other populations at a low enough level. The reason that such control may be needed is sometimes that humans have reduced the number of other top predators in the area.

Another way that we disturb ecosystems is by the introduction of new species or new variants of species into ecosystems. Sometimes we do this on purpose, as with the breeding or genetic modification (GMOs) of crops, and sometimes, it is a result of our careless transport of goods around the world, as with species that come with ballast water in ships. Some of these species may thrive in their new environment and become what we call invasive species. These are often r-strategists. Examples are unfortunately plentiful. In Sweden, a few of the current threats from invasive species are from the Iberian slug (*Arion vulgaris*), which was introduced unintentionally in the 1970s and thrives in particular on farmland and in gardens, and the mink (*Neovison vison*), which came in the 1960s when it was popular to breed them for their fur. The mink attacks the population of ground-nesting birds in the archipelagos. On the other side of the planet, the cane toad (*Bufo marinus*) is an example of an invasive species introduced with the best intentions (biological control of sugarcane pests) but which is impacting ecosystems all around the northern and eastern regions of Australia. This robust, poisonous r-strategist has spread to national parks, driving down the numbers of its potential predators, with wider population ecological effects.

2.6 Review Questions

1. Describe how each of the four environmental compartments is involved in any one of the major biogeochemical cycles, e.g. the carbon cycle.
2. Explain the reasons for the temperature profiles of the troposphere and the strato-sphere and what this means for the exchange of air between these two atmospheric layers.
3. Explain the natural greenhouse effect, and what impact it has on the average surface temperature on Earth.

4. Explain what different types of processes are involved in generating the equilibrium concentration of ozone in the ozone layer.

5. What different mechanisms drive the circulation of water in the oceans?

6. What are typical relations between energy contents in different trophic levels in a food chain? For an omnivore that can choose which trophic levels to consume from and is concerned about feeding the whole population, what does this tell us about what trophic level to consume from?

7. What are ecosystem services? Give examples of the four types of ecosystem services.

8. How large is the current ecological footprint of mankind and how is it possible that the carrying capacity of the planet can be transgressed?

9. According to the planetary boundaries concept and the researchers behind it, what Earth systems are currently at increasing or high risk and what are the reasons behind this?

3   Impacts of Chemical Pollution

3.1 Some Definitions

A 'pollutant' (or 'pollution') is a substance which is above its normal concentration range in a particular part of the environment. For example, the natural concentration of nitrogen in some rich soils would be regarded as pollution if it was obtained by using fertiliser in another environment with naturally low nitrogen levels. On the other hand, any trace level of perfluorooctanoic acid in soil, or the presence of plastic bags in the ocean, is pollution, because the natural abundance of these things is zero. This is not to say that a natural, unpolluted environment is always supportive of human life. Some authors use a different definition of pollution and like to think of volcanic sulphur emissions, volatile organic carbon emitted by plants to the air and some other chemical emissions as 'natural pollution', but these are normative, anthropocentric judgements. We might as well say the Antarctic is polluted with snow.

Pollutants may give rise to effects on a *local, regional* or *global* level, depending on how they are distributed and transformed in the environment and by which mechanisms they exert an impact on humans and the environment. Local effects take place in the environment close to the point of emission, in the direct surroundings. Such pollutants do not easily move in the environment and are typically emitted to soil, end up quickly in soil or sediments, or are short-lived. Regional effects are created further away from the emissions' source, either because emissions are moved away from the source by winds or water currents, potentially to more sensitive environments, or because emissions are the results of mechanisms that need time or certain conditions to happen. Global effects are the ones that are in principle independent of the location of the emission source. Pollutants that give rise to these will disperse and will affect the whole Earth system regardless of where they are emitted. The effects may, however, be very different in different parts of the world.

Further, pollutants may be *primary* or *secondary*, depending on whether they are the actual substances emitted or are instead the products of chemical transformation processes on primary pollutants in the environment. We often discuss emissions also in terms of whether they are *point emissions* with a few large sources or *diffuse emissions* with many small sources that are often less intentional and more difficult to control.

More details on the modelling of dispersion and on toxicology are provided in Chapters 4 and 5. In this chapter, the focus is on giving an understanding of chemical pollution in general and the different types of impacts to which it gives rise.

3.2 Acidification of Water and Soil

For a long time, the term *acidification* in these contexts referred to acidification of freshwater and soil by 'acid rain', but in recent years, we have also started to talk about other ways in which soils are acidified and about ocean acidification. This section of the book describes all these types.

Acidification of freshwater and soil is the environmental problem that led to the first large international meeting on environmental pollution, organised by the United Nations in 1972 in Stockholm. It is therefore reasonable to start with discussing this problem. Although it is not in itself a complicated issue since the pollutants have clear and understandable effects, the transboundary transport of these pollutants complicates the discussion of effective countermeasures. In fact, the countries that have experienced the largest impacts have only contributed a minor fraction of the pollution that has caused the problem. This is thus a *regional* environmental effect.

The usual causes of acidification of freshwater and soil in a North American or European context are emissions to air of acidifying substances containing sulphur and nitrogen, primarily sulphur dioxide (SO_2), nitrogen oxides (NO_x) and ammonia (NH_3). The sources of these acidifying substances are primarily traffic (land- and sea-based), industry (in particular coal burning) and agriculture. Most combustion processes generate NO_x and, if the fuel contains sulphur, also SO_2. These compounds are chemically transformed after emission, creating sulphuric and nitric acid, and are spread by winds and deposited by wet or dry deposition on soil and in water; see Figure 3.1. The application of fertilisers containing ammonium (NH_4^+) releases H^+ ions during the nitrification reaction, which forms NO_3^-. If ammonia (NH_3) is released into the air, it will be transformed into ammonium in the air and deposited with precipitation on soil and in water.

The normal pH of rain is about 5.6. It is slightly acidic because carbon dioxide (CO_2) dissolves into water, forming carbonic acid. Since 'acid rain' can additionally contain sulphuric and/or nitric acid, it has a lower pH, commonly around 4, but even lower values have not been uncommon. However, soil and surface water often have a certain buffering capacity, especially in areas where the lithosphere is rich in magnesium and potassium oxides and carbonates. This alkalinity provides resistance to acidification because it neutralises deposited acids. For this reason, some areas are not very sensitive to acidification while others are very sensitive. In fact, liming (usually the addition of ground $CaCO_3$) is a way to mitigate the effects of acidification in aquatic environments and soils used for forestry and agriculture.

When the pH of the soil or water decreases, this creates problems for organisms sensitive to the pH level; see Table 3.1 for examples of critical pH levels for some aquatic organisms. Sometimes, the organisms themselves are not very sensitive, but the organisms they feed on are. One example is the frog in Table 3.1 that is not very sensitive, but it primarily feeds on the mayfly in some ecosystems, which is an organism that is much more sensitive to pH levels.

The acidification itself is an issue as this directly affects some organisms, but secondary problems also arise. Of great concern is the ion exchange process that takes

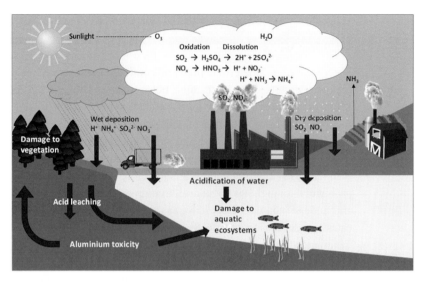

Figure 3.1   The mechanisms behind freshwater and soil acidification

Table 3.1 Critical pH Levels for Some Aquatic Organisms (US EPA, 2018)

Animal	Critical pH level
Snails	6.0
Clams	6.0
Bass	5.5
Crayfish	5.5
Mayfly	5.5
Trout	5.0
Salamanders	5.0
Perch	4.5
Frogs	4.0

place when the acidity reaches a certain level in soil. The hydrogen ions (H^+) will then replace cations on the organic material in the soil, thereby releasing for example aluminium cations into moisture in the soil, which may travel via groundwater to rivers and nearby lakes. In these aquatic ecosystems, these aluminium ions can give rise to toxic effects in e.g. fish, by clogging their gills. Nutrients also leave the soil in the same way as aluminium, potentially leaving the soil nutrient-deficient and adding excess nutrients to receiving waters.

As mentioned earlier, soils and freshwater in Scandinavia were among the first to be affected by acidification. This is due to the combination of high acid deposition levels and the low buffering capacity of soil and water. In parallel to negotiations for emissions reduction in source countries, these countries started liming the most heavily affected areas, with good effects. However, liming has to continue as long as acid

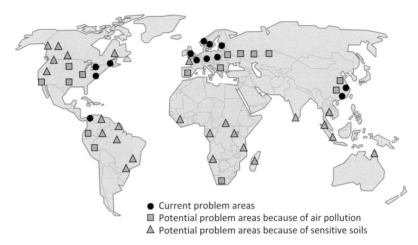

● Current problem areas
☐ Potential problem areas because of air pollution
△ Potential problem areas because of sensitive soils

Figure 3.2   Potential problem areas for freshwater and soil acidification

deposition is high. Today, emissions have been reduced dramatically, in particular for SO_2, but there is still need for liming in some areas. According to the Swedish Agency for Marine and Water Management, the counties in Sweden reported, for 2016, that about 100,000 metric tonnes of lime were spread on more than 2,700 lakes and more than 9,000 km of rivers and streams, to reduce acidification damage (HOV, 2018). This is a reduction by 45% since the early 2000s, but shows that the problem is not yet solved. Potential problem areas exist all over the world; see Figure 3.2. Current problem areas include the southern provinces of China, where strong economic growth has intensified coal burning over time and led to increased emissions of SO_2.

In many parts of the world, emissions of acidifying substances have been greatly reduced since the 1970s, because of effective national legislation and multilateral agreements; see further details on this in Chapter 8. The statistical office of the European Union (Eurostat) reports that emissions of acidifying gases in the European Union decreased significantly between 1990 and 2014, with the biggest fall for SOx emissions of almost 90% (mostly within energy production and distribution), followed by NOx emissions of 55% (mostly related to road transport). Ammonia emissions decreased by 24% (partly due to better manure management), and therefore since 2007, ammonia has been the most important acidifying gaseous emission in Europe (EEA, 2017). A large source of SOx today is international shipping, and, as land-based sources have now been greatly reduced, the proportion due to shipping is increasing and is likely to be targeted in future regulations.

In tropical, coastal locations, acidification of soils may be more of a land use management issue than related to emissions. Some soils in these locations have a naturally high level of iron sulphide (FeS_2 – 'pyrite'). When such soil is below the water table, oxygen transfer is relatively limited, but when farmers drain it prior to planting sugarcane or other agricultural activities, air penetrates the soil matrix. When this happens, the pyrite reacts with oxygen in the air and sulphuric acid is formed. At its simplest, this is shown in the equation:

$$2\ FeS_2 + 9\ O_2 + 4\ H_2O \rightarrow 8\ H^+ + 4\ SO_4^{2-} + 2\ Fe(OH)_3 \qquad (3.1)$$

Other agricultural practices also cause acidification. These include: (1) the addition of nitrogen, whether in artificial fertilisers and manure or by planting legumes; (2) the removal of agricultural products by grazing or cropping; and (3) practices that increase the amount of organic matter in the soil. The first point is that if nitrate leaches away accompanied by charge-balancing cations (e.g. Ca^+, K^+ etc.), then the hydrogen ions left behind cause acidification. Also, ammonium can react with hydroxide ions (removing them from the pH balance), making water and ammonia, which volatilises. The second point is that just growing plants and removing them can change soil pH as plants often take up more cations than anions. To compensate for the charge imbalance at the roots, they release H^+ ions into the soil. The last point is a two-edged sword: organic matter can act as an acidifying substrate or as a buffer against pH change, depending on the soil type. Acidifying processes are counteracted by the inputs of alkali into the system (via purchases of additional livestock, some fertilisers, lime and other soil ameliorants).

More recently, ocean acidification has been added to the list of acidification problems. The mechanism behind this, however, differs from land and freshwater acidification as it is due to the dissolution of atmospheric carbon dioxide in the oceans and thus related to the increase in atmospheric concentrations of carbon dioxide. In terms of abatement, it therefore has to be addressed together with climate change rather than with land and freshwater acidification. Ocean acidification results in more acidic seawater, which is problematic for some marine organisms. This affects for example organisms that depend on calcification for generation of shells and plates (e.g. prawns, foraminifera). However, resilience to ocean acidification has been shown to be larger than first expected for some organisms. In colder water, more carbon dioxide can dissolve and hence the pH can become lowest in these areas.

3.3 Eutrophication of Water and Soil

Eutrophication, or rather, hypertrophication, is the excessive addition of nutrients to an ecosystem, often a water body. This results in the excessive growth of plants and/ or algae, which may in turn give rise to several different unwanted effects, like oxygen depletion in the water body and toxic algal blooms.

The phenomenon of eutrophication is natural in most lakes and is part of the ecological succession or ageing process of lake ecosystems. This process of building up nutrient levels, sediments and productivity would normally take thousands of years, but with the anthropogenic addition of nutrients, this is a process that can instead take only decades. In a lacustrine (lake) ecosystem, phosphorus is normally the growth-limiting nutrient, so addition of phosphate from agricultural run-off or sewage speeds up the eutrophication process. In marine and terrestrial (land) ecosystems, however, nitrogen is often the growth-limiting nutrient, and nitrate in water and NOx and ammonia in air are therefore often important contributors to eutrophication in such ecosystems.

In a lake ecosystem, the nutrient addition (primarily phosphate) leads to excessive algal growth; see Figure 3.3. Above the thermocline, oxygen levels are kept high by

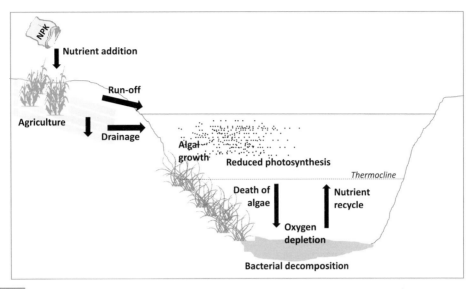

Figure 3.3 Eutrophication in a lake

diffusion from the atmosphere, oxygenated water inflows, photosynthetic algae and mixing processes. When growth is excessive, visibility is reduced and sunlight can only penetrate to a certain level. Dead algae fall to the bottom of the lake and create anoxic conditions as oxygen provision is low in these areas. In the anoxic sediments, anaerobic decomposition causes nutrients in the sediments to be recycled, further aggravating the nutrient excess. Typical for eutrophic lakes is that biodiversity is decreased and that the water may have odour and taste issues.

In marine ecosystems, similar processes occur as a result of eutrophication, leading to algal blooms and hypoxia. Episodic cyanobacterial blooms can heavily affect water quality and the toxins produced affect marine life and people swimming in the water. The socioeconomic impacts are considerable as human health, fishery, tourism and recreation are affected. The monitoring and management of the blooms are also in themselves costly.

Organic (i.e. carbon-based) pollution gives rise to similar problems as the addition of nutrients because the organic compounds become substrates for microorganism growth. The organic material may also directly cause poor visibility, reducing light available for photosynthetic organisms.

One marine area that has experienced particular problems is the Baltic Sea. The nutrients that are to blame are nitrogen and phosphorus. One important reason for this is run-off from the heavily agricultural regions along the coasts and various rivers that end up in the Baltic Sea in e.g. Sweden, Finland, Poland and the Baltic states. Before wastewater collection and wastewater treatment plants were built, sewage was also an important source of nutrients in the Baltic Sea. Nowadays, it is estimated that about 80% of the nutrients inputs come from land-based activities, including sewage, industrial and municipal waste and agricultural run-off, and that the rest is from NOx, emitted primarily from combustion processes such as traffic and the energy sector, or from industry. Fifty per cent more phosphorus enters the Baltic

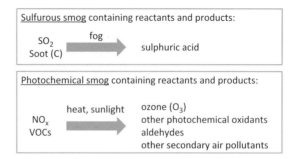

Figure 3.4   Types of smog

Sea every year, compared with 100 years ago. It accumulates in the sediments and can be resuspended by anoxic conditions at the surface of the sediments. This natural nutrient recycling sends it back to the overlying water where it contributes to algal growth. Another factor contributing to eutrophication of this water body is the slow exchange of water with other parts of the ocean through the narrow strait of Öresund. While oceans typically contain 35 g salt per kg seawater, the surface water in most of the Baltic Sea is brackish (0.5–30 g/kg) and at its northern end, the water is not salty at all and is therefore ecologically closer to a lacustrine than a marine environment. This is due to large river inflows and a strong, stratifying halocline in the central Baltic Sea.

3.4 Particles, Smog and Tropospheric Ozone

On a December morning in 1952, Londoners woke up to a five-day-long air pollution episode that has been known as the Great Smog of London or the Big Smoke. The phenomenon was a combination of weather circumstances and severe air pollution. Windless and stagnant conditions (due to temperature inversion) and heavy fog and accumulated smoke and other pollutants, primarily from domestic coal burning in the city, led to what is today known as *smog* ('smoke fog') of the *London type*, or *sulphurous, industrial* or *grey-air smog*. As shown in Figure 3.4, such smog contains sulphur dioxide, droplets of sulphuric acid and solid particles (e.g. unburned carbon – soot).

The smog caused major disruption in city life by reducing visibility and affecting health for many people. It has been estimated that about 12,000 fatalities occurred as a result of the smog, both from reduced visibility so that people were injured or fell into the river and also because of respiratory problems. This episode is known as the worst in the history of the UK, but poor air quality had been an issue at least since the 1600s. What is particularly interesting about this event is that it also had a large impact on public awareness and even led to the promulgation of the (UK) Clean Air Act in 1956. This type of smog remains a problem in urban areas of China, India, Ukraine and some Eastern European countries due to inadequate pollution control.

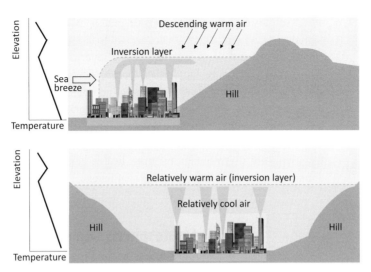

Figure 3.5 Geographic conditions that lead to temperature inversions in Los Angeles (upper figure) and London (lower figure)

Another type of smog is referred to as *Los Angeles-type, photochemical* or *brown-air smog*. The composition is different, and so are the mechanisms by which it forms. What is common with the London-type smog is that it forms under stagnant air conditions during temperature inversions where air pollutants can build up to considerable concentrations. Both these cities, London and Los Angeles, have geographical conditions that make them prone to develop temperature inversions; see Figure 3.5. In London, the whole city is low land with hills surrounding it, forming a pot that can trap air pollution; in Los Angeles, the hills that exist at some distance from the coastal city area tend to trap polluted layers during the sea breeze.

The Los Angeles-type smog contains photochemical oxidants that are secondary pollutants, formed in atmospheric reactions from traffic emissions. The presence of solar radiation is important, and therefore this problem is worse on sunny days with heavy traffic. Photochemical oxidants are the products of reactions between NOx and a wide variety of volatile organic compounds (VOCs). The most common photochemical oxidants are ozone (O_3), peroxyacetyl nitrate (PAN) and hydrogen peroxide (H_2O_2). Impacts on the natural environment are mainly due to elevated concentrations of O_3, which has toxic effects on both plants and humans. The mechanism can be simply described as shown in Figure 3.4.

Many of the constituents of smog are in themselves problematic pollutants also when they do not occur as a constituent of smog. Particles in air are primarily natural (about 60%) in the form of dust (from sea salt or from wild fires) and the rest have human sources from ploughing fields, road construction, unpaved roads, tobacco smoke, coal burning, industrial plants and motor vehicles. Inhalable particulate matter (PM) is often divided into fine, PM10 (< 10 μm), and ultrafine (PM2.5). Particulate matter can have a variety of effects, e.g. it will irritate the nose and throat, damage lungs, aggravate asthma and bronchitis, shorten life, reduce visibility, corrode metals and discolour clothes and paints (see Section 5.1.1.6 for further discussion). It can also contain toxic substances, such as polycyclic aromatic hydrocarbons (PAHs) that originate from the same sources (motor vehicle exhaust, cigarette smoke, wood smoke or fumes from asphalt roads), which adhere to the surfaces of the particles. Several of the PAHs are known to cause cancer.

The so-called Asian brown cloud is a layer of air pollution that covers parts of South Asia, including the Indian Ocean, every year typically between January and March. The cloud contains particles and other pollutants from combustion (e.g. wood fires, cars, biomass burning and factories) and industrial processes with incomplete burning. The reason for its episodic appearance is that during some parts of the year, there is no rain to wash the pollutants from the air. The effects of this cloud are diverse, ranging from health issues and reduced photosynthesis to climate change on both local and global levels.

3.5 Stratospheric Ozone Depletion

In the 1930s, Sydney Chapman proposed a model for the formation and depletion of ozone in the stratospheric ozone layer. He suggested an equilibrium concentration of ozone, resulting primarily from photolytic formation (Equation 3.2) and photolytic decomposition (Equation 3.4), effected by ultraviolet radiation of two different wavelengths in the UVB region; see Equations 3.2–3.5.

$$O_2 + UV\ (\lambda < 200\ nm) \rightarrow 2\ O\cdot \tag{3.2}$$

$$O_2 + O\cdot \rightarrow O_3 \tag{3.3}$$

$$O_3 + UV\ (\lambda = 200\text{–}300\ nm) \rightarrow O_2 + O\cdot \tag{3.4}$$

$$O\cdot + O_3 \rightarrow 2\ O_2 \tag{3.5}$$

However, the concentrations that Chapman predicted were higher than what could later be measured. In the late 1960s, Paul Crutzen added important understanding of the chemistry of the ozone layer by describing how certain molecules act as catalysts of ozone depletion; see Equations 3.6–3.8. The predicted concentrations using this more refined model were now in the same range as measured values.

$$X\cdot + O_3 \rightarrow XO + O_2 \tag{3.6}$$

$$XO + O_3 \rightarrow X\cdot + 2\ O_2 \tag{3.7}$$

$$XO + O \rightarrow X\cdot + O_2 \tag{3.8}$$

$X\cdot = NO\cdot$, directly from aeroplanes or indirectly from photolytic decomposition of N_2O; $OH\cdot$ from photolytic decomposition of H_2O; $Cl\cdot$ from photolytic decomposition of freons; $Br\cdot$ from photolytic decomposition of halons

The most important such catalyst that acts as a natural regulator of the ozone concentration in the stratosphere is nitrogen monoxide ($NO\cdot$) that originates from nitrous oxide (N_2O) that is emitted from biological processes in soil and in water at the surface of the Earth. Nitrous oxide is so stable that it can reach the stratosphere and even slowly penetrate the temperature inversion barrier between the troposphere and the stratosphere. Once up in the stratosphere, the ultraviolet radiation will split the nitrous oxide into nitrogen monoxide that will start acting as a catalyst for ozone depletion. Another chemical species that may act as such a catalyst and that may be present in low amounts in the stratosphere is hydroxyl radicals (OH·) from the splitting of water molecules (H_2O). A catalyst makes a reaction happen without being consumed. If you look at Equations 3.6–3.8, you can see that the catalyst X· is oxidised but then reduced in other reactions so that it can enter into a new cycle of depletion. Eventually, all such compounds or elements will meet their final destiny – a sink – that renders them harmless by other atmospheric reactions. Crutzen accused aeroplanes flying at high altitudes of directly emitting nitrogen oxides into the sensitive stratosphere; he thereby caused the building of supersonic aircraft fleets that were being planned at the time to slow down.

A few years later (1974), Crutzen, together with Sherwood Rowland and Mario Molina, published new and even more alarming findings that described how chlorine and bromine radicals in the stratosphere act as very potent catalysts of ozone depletion. The reaction is exactly the same as for nitrogen monoxide and hydroxyl radicals, but the atmospheric sinks for chlorine and bromine radicals are fewer and they will therefore participate in many more depletion cycles, potentially tens of thousands, before being removed. Chlorine and bromine that reach the stratosphere primarily come from the use of some very stable industrial chemicals, the so-called freons and halons. These compounds are fully halogenated hydrocarbons: freons are typically chlorofluorocarbons (CFCs) and halons are often also brominated. Species that are not fully halogenated also exist and contain some hydrogen that makes the molecule less stable and less potent. The CFCs, in particular the ones that contain high levels of fluorine, are also very strong greenhouse gases (see Section 3.6). Freons have been widely used as solvents, for example in dry cleaning, as propellants in spray cans, as refrigerants in the cooling system of refrigerators and air conditioners, and as blowing agents and insulating gases in foam insulation. Halons have primarily been used in fire extinguishers. These chemicals do not have a colour or a smell, have a very low toxicity and have excellent chemical properties for the purposes for which they were used, and therefore seemed ideal before this unexpected environmental impact was discovered.

A healthy ozone layer contains about 300 Dobson units of ozone. This means that the thickness of the ozone layer that can be found in a column drawn from the surface of the Earth and up through the whole atmosphere is about 3 mm, if the ozone was concentrated and at standard temperature and pressure (STP). However, in reality, the concentrations vary with season and latitude. Formation of ozone is strongest where sunlight is strongest, i.e. at the equator. The photolytic depletion is also strongest over the equator. But winds transport ozone-rich and hot air upwards and towards the poles, where it becomes colder and sinks and where photolytic depletion is slower. This leads to an accumulation of ozone towards the poles and lower concentrations above the equator.

Increased stratospheric ozone depletion due to emissions of CFCs, in particular, has led to a global decline of stratospheric ozone of on average 7%. To add to this, we also see a seasonal decline of ozone over the Antarctic that is the effect of the same emissions but a different mechanism, as is explained in what follows.

In 1985, a British research team discovered worryingly low concentrations of ozone in the stratosphere above Halley Bay in the Antarctic. Concentrations were lower than half of what is normal; in fact, at some altitudes, ozone was almost completely depleted. These low concentrations were seen in September and October, and after that this 'ozone hole' phenomenon seemed to disappear until the next spring, when it reappeared. The mechanism behind this ozone hole is sometimes referred to as *heterogeneous catalytic ozone depletion*, because of the involvement of certain particles. At wintertime, the air above the Antarctic becomes extremely cold because of a certain weather phenomenon that creates a large vortex of air above the Antarctic with minimal exchange of air with lower latitudes. In the darkness that prevails in wintertime at such high latitudes, temperatures fall very low in the stratosphere. Temperatures may be as low as $-80°C$, and under such conditions, certain ice particles containing nitric acid form. Also, sulphur-containing aerosols are present. The surfaces of these particles catalyse the formation of chlorine gas (Cl_2) from CFCs that are present in the stratosphere. Hence, chlorine gas is accumulated over the winter and when sunlight returns in spring, in September, the chlorine gas is easily photolytically decomposed to chlorine radicals. These enter into catalytic ozone depletion cycles as earlier explained, and ozone is thereby almost completely depleted in some layers of the stratosphere above the whole Antarctic continent during these episodes. When sunlight returns, it also leads to a weakening of the vortex of air, and after a while it is dissolved and the air is spread northwards over a larger area. Then, the ozone hole seems to temporarily heal and lower concentrations of ozone can be found also above Australia and New Zealand for a while. This heterogeneous ozone depletion also contributes to the global decline.

The energy required to drive the reaction in Equation 3.4 is what effectively stops harmful UVB from reaching different life forms in lower levels of the troposphere, which it would otherwise threaten. When concentrations of ozone are lower than normal in the stratosphere, more UVB from the sun will penetrate the atmosphere and reach sensitive organisms at the surface of the Earth and in the upper levels of the oceans. Effects on human health include eye damage (snow blindness and cataracts), skin damage (sunburn and skin cancer) and immune system suppression. Ecosystems are damaged in various ways. Some organisms are very sensitive (such as leguminous plants and phytoplankton) whereas some are less sensitive (such as grasses). This reduces crop yields, suppresses oxygen production in the oceans and changes the competition situation between species. Other types of effects include damage to some industrial materials (e.g. plastics).

The same year as the ozone hole was discovered, discussions started on how to reduce emissions of stratospheric ozone depleters. These discussions were very successful, as is discussed in Chapter 8.

3.6 Climate Change

As discussed in Section 2.1.1, certain greenhouse gases in the atmosphere capture heat radiation and thereby delay the departure of energy from the Earth system, increasing the average temperature. The Swedish physicist Svante Arrhenius

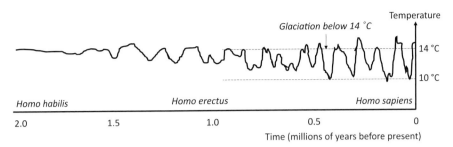

Figure 3.6 Global average surface temperature variations in the Quaternary (Pleistocene)

explained this as early as 1896. He made calculations that showed that a doubling of the carbon dioxide concentration in the atmosphere would lead to an increase in the average temperature on Earth of about 6°C, of which 2°C would be caused by the increased level of water vapour in the atmosphere that would be a feedback effect of the warming. It would, however, take many years before the enhanced greenhouse effect (i.e. beyond the natural greenhouse effect) and the resulting temperature and climate changes would become a general concern and become matters for political discussion. And although Arrhenius' calculations indicated a certain temperature increase, we now know that the climate system is more complicated and encompasses more interacting mechanisms than Arrhenius included in his model.

The average temperature on Earth is determined primarily by how much solar radiation enters the Earth system. This is, in turn, determined by the intensity of the sun and the position of the Earth relative to the sun. As both solar activity and Earth's orbital patterns vary over time in a cyclical way, there is also a natural cyclical variation in the average temperature on Earth. Geologists characterise the Quaternary Period as strongly influenced by such patterns, resulting in a cyclic temperature variation; see Figure 3.6, with resulting growth and decay of continental ice sheets. The last glacial period ended about 12,000 years ago, and it is during the stable and comfortable temperature (of more than 15°C) of this last interglacial – the Holocene – that human society evolved to what it is today, with agricultural activities starting around the end of the glacial; see Figure 3.7. We should now slowly be heading towards a new cooler period. This is, however, not what we are currently observing. Rather, we are worried that human activities are disturbing the climate system on Earth in a way that may lead to severe consequences.

However, the climate system is quite complex. The extent to which our planet reflects solar radiation is called the *albedo* of the Earth. Clouds and particles in the atmosphere reflect incoming solar radiation, as do optically bright surfaces like ice and deserts. An increased albedo will have a cooling effect as it will stop some solar radiation from being absorbed in the system and re-emitted as heat.

The concentration of greenhouse gases in the atmosphere determines how strongly heat is trapped in the system. Water in vapour form is actually the greenhouse gas that generates most of the natural greenhouse effect. The concentration of water in different parts of the atmosphere and its exact form is a result of many different parameters and mechanisms, for example temperature, winds and the presence of particles acting as condensation nuclei. Therefore the extent to which water will be

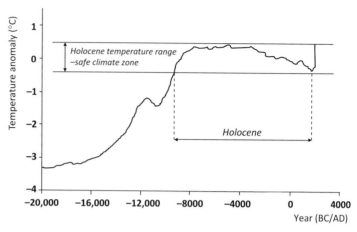

Figure 3.7 Global average surface temperature variations in the Holocene

present as vapour, trapping heat, or as droplets, reflecting incoming solar radiation, is difficult to predict in detail. We normally do not consider emissions of water to the atmosphere to increase the greenhouse effect as the presence and form of water in the atmosphere is primarily a result of mechanisms other than the release from human activities.

Carbon dioxide, however, is a greenhouse gas that is increasing in concentration due to human activities. Large, natural flows of carbon dioxide to and from the atmosphere are the result of the photosynthesis and respiration of green plants and of a chemical balance with carbonates in the oceans, as outlined for the carbon cycle in Section 2.2.2. However, when humans disturb these balances by for example deforestation, the concentration increases. Also, when humans introduce carbon that is normally not in circulation in the carbon cycle, for example by burning fossil fuels or by the production of cement, the concentration in the atmosphere will also increase. From pre-industrial times, the atmospheric concentration of carbon dioxide has increased from about 270 ppm to more than 400 ppm; see Figure 3.8. Note that there is an annual cycle for the carbon dioxide concentration on Earth with a lower concentration when the northern hemisphere, which has the largest land masses and therefore the largest amount of land biomass, has its summer from June to August, because the largest amounts of carbon are then tied up in that biomass.

Methane is another greenhouse gas that is increasing due to human activities. Methane is naturally formed during anaerobic decomposition of organic matter, and therefore landfills and rice paddies and other wetlands generate large amounts. Also, when humans extract fossil fuels from the lithosphere, large amounts of methane escape to the atmosphere. Another large source is cattle, as these ruminants generate large amounts of methane when digesting their food. From pre-industrial atmospheric concentrations of about 722 ppb, the concentration is now about 1,900 ppb. This greenhouse gas is stronger than carbon dioxide, i.e. it has a greater ability to capture heat energy. According to the IPCC (2014b), it is 28 times more effective than carbon dioxide on a mass basis (in a 100-year perspective).

Nitrous oxide (N_2O) (laughing gas) is an even stronger greenhouse gas, 265 times stronger than carbon dioxide per kg (in a 100-year perspective, according to the IPCC,

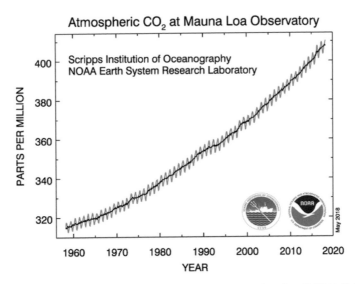

Figure 3.8 Carbon dioxide concentration in the atmosphere as measured at Mauna Loa, Hawaii (NOAA Global Monitoring Division, 2018; automatically generated graph)

2014b). From pre-industrial levels of 270 ppb, the atmospheric concentration has now risen to more than 330 ppb. Nitrous oxide forms naturally during nitrification and denitrification processes in soil and water. When we increase the availability of nitrogen by using mineral fertilisers in agriculture, the rate of formation increases. Further, catalytic converters in cars and efforts to reduce NOx emissions from combustion have generated increased emissions of laughing gas due to the reducing conditions.

Some halogenated substances are also strong greenhouse gases, in particular the freons and other fluorinated compounds. These are thousands of times stronger than carbon dioxide, but their use and emissions have been limited compared with other greenhouse gases, partly because of their regulation as causes of stratospheric ozone depletion.

These greenhouse gases are all so-called well-mixed greenhouse gases because their relatively long lifespans allow them to be evenly distributed around the planet. Thus they have a global impact. The contributions of the different major well-mixed greenhouse gases to the increase in the greenhouse effect and the major emitting sectors are provided in Figure 3.9.

However, there are also near-term climate forcers that have regional and near-term effects. This includes some of the methane, tropospheric ozone and aerosols (or their precursors) and some halogenated gases.

All the greenhouse gases have the ability to change Earth's radiation balance as they temporarily hinder outgoing heat radiation. Aerosols and particles in the atmosphere have a counteracting effect as they hinder incoming solar radiation. Therefore, volcanic eruptions and the presence of clouds will have an impact. When the surface of the Earth is changed, this can have similar effects if the albedo of the surface changes. In general, dark and rough surfaces capture more incoming solar radiation while more is reflected away when surfaces are bright or smooth. Ice, for example,

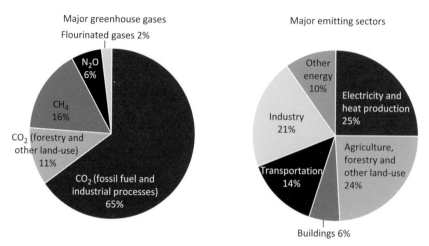

Figure 3.9 Major global greenhouse gases and major global emitting sectors (IPCC, 2014b; based on global emissions from 2010)

reflects much of the solar radiation while open water captures it, and the melting of polar ice caps will thereby lead to a positive feedback mechanism that induces even more warming.

As the radiation balance changes so that more energy is contained in the atmosphere and the hydrosphere, the average surface temperature on Earth increases. The observation that this is causing dramatic changes to natural ecosystems is one of the major reasons that some researchers are now using the term *Anthropocene* to indicate the shift into a new geological era with industrialisation, an era in which the Earth system is dominated by the activities of people. The UN Intergovernmental Panel on Climate Change (IPCC), consisting of thousands of scientists from around the world, gathers the latest scientific facts on climate change in a report that is published every five years. The IPCC concluded in the fifth assessment report (IPCC, 2014b) that human activities have resulted in an increase of the globally averaged combined land and ocean surface temperature of 0.85°C over the period 1880 to 2012.

But an increase in the global average surface temperature does not mean that every place on Earth will be warmer. The temperature increase will alter the climate in various ways. Exactly how the climate will change in each location is even more difficult to predict than long-term weather forecasts are to make. But in general, as the hydrological cycle speeds up in a warmer climate, climate will be not only warmer but also wetter and wilder. However, some areas that are already very dry may become even drier. Further, when water gets warmer, it takes up more space, and this will lead to rising sea levels. When this is combined with a wilder climate with more frequent and more severe storms, coastal areas as well as low coral islands may be heavily affected. If the thermohaline water circulation in the world's oceans changes dramatically, this would severely impact the climate on all continents. With changing temperatures, climate zones move and flora and fauna are affected. Some diseases may move into new areas. For example, malaria is a tropical, mosquito-borne disease, but if the habitat for these mosquitos spreads away from the equator, the disease will follow. The list of potential impacts can be made very long. What is of most concern is if the whole Earth system would be forced to shift into a new climate regime, as this would greatly change the conditions for human society everywhere, with very large associated costs and mass migration of people from areas that become less habitable.

Reducing the emissions of greenhouse gases is an urgent matter, but there have also been suggestions on how to speed up the removal of atmospheric greenhouse gases or to counteract climate change in artificial ways. Some of these suggestions go under the name of geoengineering. Artificial trees that capture carbon dioxide from the atmosphere, devices that increase cloud formation and large mirrors in space that reflect solar radiation are some examples. As we have already set in motion a climate change that will continue for a long time on Earth even if we dramatically reduce all greenhouse gas emissions now, we will not only have to address future emissions but also adapt to a changing climate. International climate negotiations are mentioned in Chapter 8 of this book.

3.7 Persistent Pollutants: Organic Compounds and Heavy Metals

A Who's Who of pesticides is therefore of concern to us all. If we are going to live so intimately with these chemicals – eating and drinking them, taking them into the very marrow of our bones – we had better know something about their nature and their power.

Rachel Carson, *Silent Spring*, p. 17

This quote comes from a book that was an important eye-opener for many regarding the impacts of *persistent pollutants* on human health and the environment. It hints at our lack of knowledge about the potential consequences of living in an increasingly complex chemical cocktail. The book was published in 1962 by American biologist Rachel Carson. At the time, more and more evidence was seen on the indirect effects of our use of heavy metals and organic compounds for various purposes, for example as agricultural biocides and industrial chemicals.

An early example of an agricultural biocide is dichlorodiphenyltrichloroethane (DDT) (Figure 3.10). This insecticide was invented by Swiss chemist Paul Hermann Müller in 1939, for which he was awarded the Nobel Prize in 1948. This insecticide was highly appreciated and widely used for many years for mitigating different pests; see Figure 3.11. It was highly effective towards target organisms and was considered to have very low toxicity for other exposed organisms, including humans. However, the use of DDT was subject to strong regulations from the 1970s after it was shown to be responsible for declining populations of top predator birds.

For example, it was shown that peregrine falcons (*Falco peregrinis*) and ospreys (*Pandion haliaetus*) bioaccumulated DDT and its metabolite (breakdown product) to such high concentrations in their own bodies that their calcium metabolism was disturbed and hence their eggshells became too thin to protect their offspring.

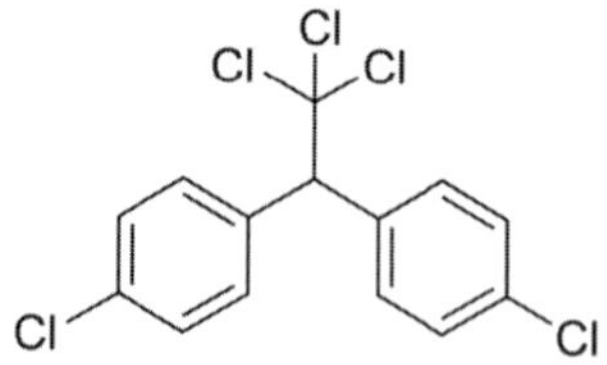

Figure 3.10 Chemical structure of the insecticide dichlorodiphenyltrichloroethane (DDT)

Figure 3.12 shows how DDT bioaccumulates in the fatty tissue of different organisms and can thereby biomagnify in the food chain and ultimately lead to very high concentrations in the upper trophic levels. When used in large amounts, DDT was found in run-off from sprayed farmland and thus in receiving waters. Plankton living in the water accumulates DDT to higher concentrations than in the surrounding water, small fish that ingest plankton to higher levels than in the plankton and so on.

Today, many other toxic effects of DDT or its metabolites have been demonstrated or are suspected. DDT is still used in some areas for fighting malaria mosquitos and in those areas, mothers' breast milk has concentrations of DDT far above acceptable levels.

Another example of a biocide that bioaccumulates is alkylmercury compounds that were used as seed dressings to control seed-borne diseases in cereal crops, a practice that started around 1924. In 1965, it was discovered in Sweden that seed-eating birds were strongly affected. In Alberta (Canada), the hunting of pheasants and partridges was prohibited due to mercury contamination which was thought to have been caused by seed dressing (Fimreite, 1970). The tragic potential for bioaccumulation and toxicity of organomercury was reported earlier in Minimata (Japan) in 1956. The so-called Minimata disease is a neurological syndrome caused by severe mercury poisoning. Minimata Bay and the Shiranui Sea were contaminated by industrial wastewater containing mercury from the Chisso Corporation's acetaldehyde plant, which grew from a small plant beginning in 1932 and continued operation all the way to 1968. Mercury, used as an industrial catalyst, was discharged as Hg (II) but was transformed by bacteria in the sediments to toxic methyl- and dimethylmercury. Concentrations of mercury of up to 2 kg per tonne of sediments were reported. This alkylmercury bioaccumulated in shellfish and fish and was eaten by the local population, resulting in mercury poisoning with symptoms like ataxia, muscle weakness, insanity and death (e.g. Kudo et al., 1998).

What is special about persistent pollutants like DDT and alkylmercury is their ability to accumulate in the environment, eventually reaching concentrations that are harmful to organisms. When an organism absorbs a substance at a rate that exceeds its ability to break down or excrete it, this is called *bioaccumulation*. This can be driven by *bioconcentration* – the accumulation of a chemical from the surrounding water – and *biomagnification*, which is bioaccumulation in the food chain (see Figure 3.12 for an example that concerns DDT). Both effects are accentuated by the persistence and the lipophilicity of chemicals, e.g. a half-life of several months or years and log K_{ow} above 5. Due to the lipophilicity, such substances are primarily found in the fatty tissue of organisms, and may, unfortunately, be secreted via lactation. Many of these substances are proven endocrine disruptors; see Chapter 5.

Figure 3.11 Examples of uses of DDT and of the perception of its impacts in the 1950s. The first example is a DDT wallpaper advert from 1947 (image courtesy of Joe Wolf, https://flic.kr/p/atcdbd), and the second example (see next page) is a photo of a US soldier that demonstrates the use of DDT-hand-spraying equipment (image courtesy of the Centers for Disease Control and Prevention [CDC])

Heavy metals other than mercury have caused concern over the years. Some key examples include lead in plumbing and paint, and in gasoline (as tetraethyl lead, which is released to air after the fuel burns), tin in antifouling boat paints (as tributyltin, intentionally slowly released to water), cadmium in mineral fertiliser

Figure 3.11   (cont.)

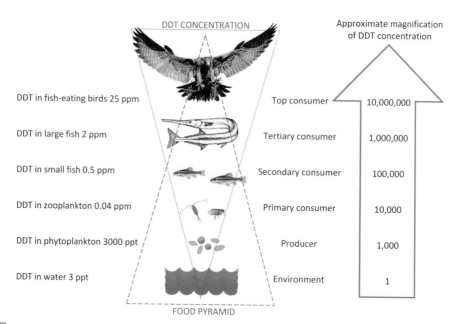

Figure 3.12   Illustration of biomagnification in the food chain for DDT

and naturally occurring arsenic in irrigation water and soil, which is taken up by crops. Lead toxicity is discussed further in Chapter 5.

There are only so many metals in the periodic table, but a list of persistent organic pollutants (POPs) of concern, other than the DDT insecticide, is potentially infinite. One such early list (created for the Stockholm Convention on Persistent Organic Pollutants – see Chapter 8) has been called the 'dirty dozen'

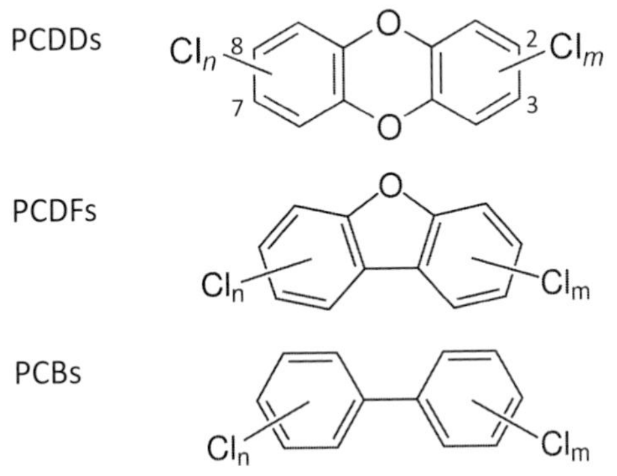

Figure 3.13   General chemical structure of the substances in the PCDD, PCDF and PCB groups

and thus contains 12 substances or groups of substances considered to be among the worst persistent pollutants. Most of this list consists of other biocides – aldrin, dieldrin, endrin, chlordane, hexachlorobenzene, toxaphene and heptachlor – but also polychlorinated biphenyls (PCBs) (a group of related industrial chemicals used in carbonless copy paper and as heat transfer or dielectric fluids in electrical apparatus) and two groups of by-products of chlorine chemistry or combustion processes – polychlorinated dibenso-p-dioxins (PCDD) ('dioxins') and poly-chlorinated dibensofurans (PCDF) ('furans'). All of the 12 items on the list are chlorinated, have aromatic structures and high toxicity (are often proven endo-crine disruptors) and are persistent and lipophilic and will therefore accumulate in the environment and biomagnify in food chains. They also have a large potential to be transported far from the emission source because they are semi-volatile and can move with winds, e.g. via the so-called global distillation effect. Also called the grasshopper effect, this is the observation that semi-volatile pollutants move with rising hot air and fall down with sinking cold and dry air to be deposited further towards the poles. With the diurnal or seasonal temperature variations they may then possibly rise again with warm, moist air and keep moving towards the poles. This effect explains why indigenous people of the Arctic have some of the highest body burdens of POPs, despite not using them in their daily lives.

It is not surprising that the biocides exhibit toxic effects also on organisms other than the target organisms. Most of the 12 items on the list are intentionally produced and are more easily controlled than the ones that are by-products or pollutants formed in many different processes, which is the case for dioxins and furans. PCBs have a history similar to DDT with toxic effects being seen in marine mammals after extensive use, and regulatory control also began in the early 1970s. The toxicity varies greatly between the 209 congeners of the PCB group; see the general structure in Figure 3.13. The dioxins and furans also contain a large number of congeners that have a varying number and position of chlorine atoms; see Figure 3.12. For the dioxins, the most toxic one is chlorine substituted at the 2, 3, 7 and 8 positions. Note the similarity in chemical structure between these substances and that of DDT in Figure 3.10.

For the persistent pollutants that have been mentioned in this section, toxic effects have been established, although some suspected toxic properties have not yet been fully established. For endocrine disruption, knowledge is particularly poor. Theo Colburn, a US pharmacist who passed away in 2014, was an important person in raising awareness and knowledge in this field. She founded the Endocrine Disruption Exchange (https://endocrinedisruption.org/), a science-based, non-profit research institute, in 2003.

Our knowledge about persistent pollutants of concern increases all the time, and it is likely that some chemicals that are not yet seen as problematic will be listed in the future. Also, new chemicals are continuously invented and taken into production, although regulations on preapproval are being introduced in many countries, e.g. in the United States and in the EU; see more information on this in Chapter 8. Nevertheless, and as regulatory work is sometimes not quick enough to prevent harm at an early enough stage, governments and NGOs have lists of emerging pollutants and observation lists that are constantly scrutinised, e.g. the SIN (Substitute it now!) list by Chemsec (The International Chemical Secretariat, an NGO; see http://chemsec.org/business-tool/sin-list/).

Chemical pollution is also discussed later in this book: in the next chapter, approaches to modelling their transport and fate in the environment are discussed; Chapter 5 discusses their potential toxic effects; and Chapter 6 introduces ways to combine transport and effect information to form a chemical risk assessment.

3.8 Review Questions

1. By which different mechanisms are fish in a lacustrine ecosystem typically affected when large amounts of acidifying substances have been released into the air in the region?
2. What does 'nutrient recycling' mean in the context of eutrophication?
3. Explain the mechanisms behind and the constituents of the two different types of smog that have been named after cities that have experienced large problems in relation to each of them: Los Angeles and London. Also explain the mechanisms behind the temperature inversions in the two different cases.
4. Why does the Earth seem to be taking one breath every year, judging by the patterns of carbon dioxide concentration in the atmosphere?
5. What are the similarities between the compounds shown in Figures 3.10 and 3.13 in terms of the chemical structure and their environmental impacts?
6. Explain the concept of biomagnification using the example of DDT.

Modelling Environmental Transport and Fate of Pollutants

4.1 Basic Concepts and Terminology

Understanding environmental fate and transport of chemicals is particularly important in relation to three other chapters in this book that deal with toxicology, risk assessment and environmental assessment of products. In toxicology, we need to know the concentrations of chemicals in the environment, since as Chapter 5 explains, the 'dose differentiates a poison and a remedy'. This chapter is about calculating those concentrations. Along with dose-response data, chemical concentrations are key inputs in quantitative chemical risk assessment. In turn, the assessment of chemical risks is one of the key issues in environmental life cycle assessment when it is applied to product design. In product design processes, policy debates and many other situations, we need to use simplified environmental models to perform quantitative chemical risk assessment on all the important emissions caused by a product or process. So experts around the world have spent worker-centuries trying to find the right balance between simple environmental models that can be reasonably well populated with data, and complicated ones that take all the issues into account (but would require many assumptions). In this chapter, we look at the basis for some simple models of chemical transport and fate. In the context of this book, the word *chemical* refers to any organic or inorganic substance, be it an elemental material or a compound, which exists as solid, liquid or gas, and may change between these phases of matter.

4.2 Transport between the 'Spheres'

You will recall from Chapter 2 how we described the planet in terms of the hydrosphere, atmosphere, lithosphere and biosphere. These terms are useful for describing not only the natural environment but also the fate of contaminants. To these natural 'spheres' we now add the 'technosphere'. We define this in terms of all of the products, structures and systems that humans have constructed – including houses, mobile phones, farms and landfills. Once chemicals have been emitted from the technosphere to one or several of the other spheres, they can move between them (see Figure 4.1).

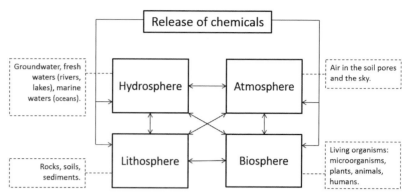

Figure 4.1   Emission of chemicals into different spheres of the environment and subsequent movement between the different spheres

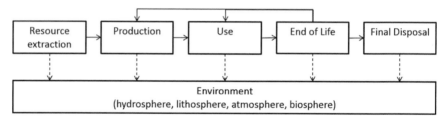

Figure 4.2   The emission of chemicals can happen anywhere in the lifetime of a product or service. Chemicals can furthermore also remain in the technosphere through recycling (end of life -> production) or reuse (end of life -> use) of products that reached their end of life

The emission of chemicals can be unintended (e.g. shipwrecks) or deliberate disposal of wastes (e.g. sewage, landfilling) and application of chemical products (e.g. fertiliser, pest control). The emission of chemicals can furthermore happen anywhere in the life cycle of a product or service (Figure 4.2). The emission of chemicals from the technosphere will generally fit the definition of 'pollution' used in the previous chapter.

4.2.1 Emission into the Hydrosphere

A direct emission of chemicals from the technosphere into the hydrosphere can for example result from the discharge of wastewater treatment plant effluent, the release of cooling water from nuclear power stations, waste dumping at sea or the emission of hydrocarbons from oil rigs, terminals or shipwrecks. Furthermore, chemicals can enter the hydrosphere through for example run-off from land, leachate from soil or atmospheric deposition.

As an example of the contaminants in one of these flows, domestic wastewater may contain various pharmaceuticals, hormones and detergents. Furthermore, storm water run-off may contain substances leached from car tyres and roofing materials, like zinc and other heavy metals, as well as organic contaminants such as nonylphenols and phthalates.

4.2.2 Emission into the Lithosphere

A direct emission of chemicals from the technosphere into the lithosphere can for example result from waste deposition at landfills, the use of pesticides or the land application of sewage sludge. Furthermore, chemicals can enter the lithosphere through for example flooding by rivers or seas, or atmospheric deposition.

One example of a pollutant discharged to the lithosphere is dichlorodiphenyltrichloroethane (DDT), a colourless and tasteless organochlorine insecticide. DDT is persistent and readily absorbed by soils and sediments, see Section 3.7. Therefore, soils and sediments can act both as contaminant sinks and as long-term contaminant sources contributing to the exposure of terrestrial and aquatic organisms to DDT.

4.2.3 Emission into the Atmosphere

A direct emission of chemicals from the technosphere into the atmosphere can occur via domestic or industrial chimneys, exhaust systems connected with internal combustion and jet engines, pesticide application using aeroplanes or the use of aerosol cans. Furthermore, chemicals can enter the atmosphere from other spheres through for instance volatilisation or wind erosion.

The combustion of wood is a potential source of heavy metals and dioxins, depending on the type of wood used and the way the wood was treated. Wood may be treated with wood preservatives containing chlorinated compounds such as DDT or Lindane (γ-HCH). The presence of chloride is key to the formation of dioxin during combustion. Aerosol cans typically use hydrocarbons to propel payloads like paints, cosmetics and personal pesticides, so they release the hydrocarbons to the atmosphere when used.

4.2.4 Emission into the Biosphere

A direct emission of chemicals from the technosphere into the biosphere can for example result from the medication of humans and animals (e.g. antibiotics, hormones, anti-inflammatory drugs) or pest control. Furthermore, organisms can take up chemicals from water, soil, air or via the food chain.

4.3 Transport Processes

To describe how chemicals can be transported within a sphere or between spheres, we need to discuss how fundamental transport and fate processes operate at the substance or even at the molecular level. In this chapter, we divide them up into food chain processes, inter-media transport processes and intra-media transport processes.

4.3.1 Transport of Chemicals by Food Chains

Chemicals can be transported along food chains. Bioaccumulation was defined in Section 3.7. *Bioavailability* or the *bioavailable fraction* is the fraction of a chemical's concentration that is effectively available for interaction with biota. For example, heavy metals in contaminated sediment may be strongly bonded with iron and manganese minerals, or even 'locked up' within the silicate structure of sediment grains. Metal in these fractions of the sediment is relatively unlikely to come into contact with food chains. On the other hand, metals weakly bonded or dissolved in the sediment pore water will more readily come into contact with sediment-dwelling organisms. To examine the bioavailability of contaminants from soils and sediments, a great deal of effort has been put into the creation of test methods that firstly extract the most bioavailable fraction, and then 'sequentially extract' increasingly less bioavailable contaminants (Tessier et al., 1979).

The operation of bioaccumulative processes will naturally mean that wherever an animal goes, it takes the chemicals it has accumulated in its body tissue with it. Migratory birds will carry persistent organic pollutants to wherever they fly and die. Another interesting example of this is the flow of nitrogen in rivers with salmon populations. It has been shown that when these fish swim and leap upstream in the coniferous forests of northern America, they bring with them significant quantities of nitrogen and phosphorus to forest ecosystems. Piscivorous birds and other animals take their catches away from the stream to get away from their competitors – bears have been observed carrying fish 800 m from streams to have a quiet meal. The birds and bears excrete some of what they eat at their feeding locations, and may leave some of the fish uneaten when they have had their fill. Then, smaller organisms rapidly incorporate it into the soil. Scientists have shown that the nitrogen from the fish is an important contribution to the total nitrogen supply that the forest receives near rivers where salmon spawn. Some trees are growing three times faster than they will if the salmon die out (Helfield & Naiman, 2001).

4.3.2 Inter-Media Transport

Inter-media transport is the transport or transfer of chemicals across phase interfaces. This includes processes like evaporation, condensation, dissolution, sublimation and adsorption. Some of the key properties determining the potential for inter-media transport are shown in Table 4.1, along with examples of the mathematical constants used to describe the property.

Henry's law says that given enough time to equilibrate, the ratio of the concentration of a chemical in aqueous solution to the concentration in the air above it is a constant that only depends on the temperature but not the concentrations, i.e.:

$$H = \frac{[A]}{p_A} \tag{4.1}$$

where H is in moles per litre per atmosphere, and where p_A is the vapour pressure of A in the air. You need to be aware that Henry's law gets 'written down different ways

Table 4.1 Phase Interfaces and Chemical Properties Controlling Inter-Media Transport

Phase interfaces	Determining physical property	Example descriptor
air-water; soil-air	vapour pressure	Henry's law constant H
water-soil; air-water	water solubility	solubility constant K_S
water-soil; soil-air	adsorption	soil organic carbon sorption coefficient K_{oc}
water-biota	lipophilicity	octanol-water partition coefficient K_{ow}

up' (i.e. sometimes the formula has the liquid phase concentration on the bottom, the inverse of Equation 4.1). So be sure to double-check the units next time you need to look up an H value, and make sure you do not invert it accidentally! Sometimes it is useful to obtain a dimensionless H, the ratio of two concentrations, so for example if we consider carbon dioxide, which has H = 0.034 mol.L^{-1}.atm^{-1}, then for transfer from air to water (at 25°C, i.e. T = 298 K):

$$H' = \frac{[A]_{water}}{[A]_{air}} = HRT = 0.034 \times 0.082 \times 298 = 0.83 \tag{4.2}$$

since R = 0.082 L.atm.K^{-1}.mol^{-1} (the universal gas constant) Again: this can be inverted, so be vigilant!

For a two-ion salt AB that dissolves into A and B, solubility is classically described by a solubility constant:

$$K_S = \frac{[A(aq)][B(aq)]}{[AB(s)]} \tag{4.3}$$

where the activity of a pure solid is, by definition, one. Thus $K_S = [A(aq)][B(aq)]$.

Sorption of a chemical from water onto a solid can at its simplest be described by an equilibrium partition coefficient as:

$$K_d = \frac{S}{C} \tag{4.4}$$

with units of volume per mass, where C is the concentration in the liquid and S is the concentration in the solid phase. This can be good enough for some situations with dilute chemicals, but in many applications, for example drinking water treatment systems, the relationship is not actually linear over the concentration range of interest, and the Freundlich equation is more realistic:

$$x = mKC^{1/n} \tag{4.5}$$

Here x is the mass of the chemical adsorbed [mg], m is the mass of adsorbent [g], C is the concentration (mg/L) of the chemical in solution at equilibrium with the adsorbent, and the pronumerals n [dimensionless] and K [(mg/g)(L/mg)$^{1/n}$] are empirical constants. We can call 1/n the *Freundlich isotherm intensity parameter* and the K the *Freundlich isotherm capacity parameter*. Curves of this kind look like the example given in Figure 4.3. It is also possible to develop partition coefficients for soil-air partitioning (Thoma et al., 1999).

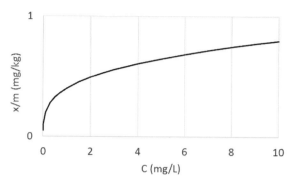

Figure 4.3 Example Freundlich sorption isotherm

One of the partition coefficients that most concerns analysts interested in hazard assessment of organic chemicals in the environment is the octanol-water partition coefficient, which is the ratio of the concentrations of a chemical in these two solutes when they are in contact with each other and at equilibrium:

$$K_{ow} = \frac{[A]_{oct}}{[A]_{water}} \tag{4.6}$$

This is a useful proxy for the extent to which a chemical will accumulate in biota, and it is used in many regulatory and legal settings. It has been suggested that chemicals with $\log K_{ow} > 4.3$ have a high potential to bioaccumulate, whereas when $\log K_{ow} < 3.5$, a chemical has low potential to bioaccumulate (Allen, 2001). Between these values, the potential to bioaccumulate is moderate.

The most accurate way to determine the K_{ow} of a chemical is to test it in a laboratory – adding it to a container with water and octanol, agitating and measuring the concentration of the chemical in both phases after equilibration. But such measurements take time, and we may want to estimate the K_{ow} of a chemical that we are planning to synthesise – one that does not exist yet. For such challenges, scientists have modelled structure-activity relationships to help predict K_{ow} and many other important chemical characteristics. Meylan and Howard (1995) pioneered these methods. Using a large database of organic chemicals, they proposed a model based on the coefficients f_i in Table 4.2, the correction factors n_j in Table 4.3 and Table 4.4 and the following equation:

$$logK_{ow} = \sum f_i n_i + \sum c_j n_j + 0.229 \tag{4.7}$$

where n_i and n_j are the number of certain structural features in a molecule and f_i and c_j are empirical coefficients for these features. For example, consider methyl salicylate (2-hydroxybenzoic acid methyl ester) as shown in Figure 4.4. With reference to Table 4.2, we can see six aromatic carbon atoms in the benzene ring (for which $f_i = 0.294$), an esteric attachment to the ring (-C(=O)O- with $f_i = -0.7121$), a methyl group

Table 4.2 Contribution Coefficients (f_i) for Octanol-Water Partition Coefficient (Adapted from Meylan and Howard [1995] with Permission of the Publisher)

	f_i		f_i
Aromatic atoms		**Aliphatic nitrogen**	
Carbon	0.2940	-NO$_2$ (aliphatic attachment)	−0.8132
Oxygen	−0.0423	-NO$_2$ (aromatic attachment)	−0.1823
Sulphur	0.4082	> N < (+5 valence, single bonds)	−6.6000
Aromatic nitrogen		-N = C = S (aliphatic attachment)	0.5236
N = O, oxide type	−2.4729	-N = C = S (aromatic attachment)	1.3369
+5 valence type	−6.6500	-NP (phosphorus attachment)	−0.4367
at a fused ring location	−0.0001	-N (two aromatic attachments)	−0.4657
in a 5-member ring	−0.5262	-N (one aromatic attachment)	−0.9170
in a 6-member ring	−0.7324	-N(O) (nitroso, +5 valence)	−1.0000
Aliphatic carbon		-N = C (aliphatic attachment)	−0.0010
-CH$_3$ (methyl)	0.5473	-NH$_2$ (aliphatic attachment)	−1.4148
-CH$_2$-	0.4911	-NH- (aliphatic attachment)	−1.4962
-CH<	0.3614	-N< (aliphatic attachment)	−1.8323
>C<	0.2676	-N(O) nitroso	−0.1299
C (no hydrogens)	0.9723	-N = N- (azo, includes both N)	0.3541
Olefinic and acetylenic C		**Aliphatic oxygen**	
=C< (2 aromatic attachments)	−0.4186	-OH (nitrogen attachment)	−0.0427
=CH$_2$	0.5184	-OH (phosphorus attachment)	0.4750
=CH- or =C<	0.3836	-OH (olefinic attachment)	−0.8855
≡CH or ≡C-	0.1334	-OH (carbonyl attachment)	0.0
Carbonyls and thiocarbonyls		-OH (aliphatic attachment)	−1.4086
CHO- (aldehyde, aliphatic attachment)	0.9422	-OH (aromatic attachment)	−0.4802
CHO- (aldehyde, aromatic attachment)	−0.2828	=O	0.0
-C(=O)OH (acid, aliphatic attachment)	−0.6895	-O- (two aromatic attachments)	0.2923
-C(=O)OH (acid, aromatic attachment)	−0.1186	-OP (aromatic attachment)	0.5345
-NC(=O)N- (urea type)	1.0453	-OP (aliphatic attachment)	−0.0162
NC(=O)O (carbamate-type)	0.1283	-ON- (nitrogen attachment)	0.2352
NC(=O)S (thiocarbamate-type)	0.5240	-O- (carbonyl attachment)	0.0
-C(=O)O (ester, aliphatic attachment)	−0.9505	-O- (one aromatic attachment)	−0.4664
-C(=O)O (ester, aromatic attachment)	−0.7121	-O- (aliphatic attachment)	−1.2566
-C(=O)N (aliphatic attachment)	−0.5236	**Phosphorus**	
-C(=O)N (aromatic attachment)	0.1599	-P=O	−2.4239
-C(=O)S (thioester, aliphatic attachment)	−1.1000	-P=S	−0.6587
-C(=O)- (noncyclic, 2 aromatic attachments)	−0.6099		
-C(=O)- (cyclic, 2 aromatic attachments)	−0.2063		
-C(=O)- (cyclic, aromatic, olefinic attachment)	−0.5497		
-C(=O)- (olefinic attachment)	−1.2700		
-C(=O)- (aliphatic attachment)	−1.5586		
-C(=O)- (one aromatic attachment)	−0.8666		
NC(=S)N (thiourea-type)	1.2905		

Table 4.2 (Cont.)

	f_i		f_i
Cyano (-C≡N)		**Aliphatic sulfur**	
N≡CS (sulphur attachment)	0.3540	-SO$_2$ N (aromatic attachment)	−0.2079
N≡CN (nitrogen attachment)	0.3731	-SO$_2$ N (aliphatic attachment)	−0.4351
C≡NC=N	0.0562	-SO$_2$OH (sulfonic acid)	−3.1580
-C≡N (other aliphatic attachment)	−0.9218	-SO$_2$O (aliphatic attachment)	−0.7250
-C≡N (aromatic attachment)	−0.4530	-S(=O)- (one aromatic attachment)	−2.1103
Halogens		-SO$_2$- (one aromatic attachment)	−1.9775
Halogen {all} (nitrogen attachment)	0.0001	-SO$_2$- (two aromatic attachments)	−1.1500
-F (aliphatic attachment)	−0.0031	-SO$_2$- (aliphatic attachment)	−2.4292
-F (aromatic attachment)	0.2004	-S(=O)- (aliphatic attachment)	−2.5458
-Cl (aliphatic attachment)	0.3102	-S-S (disulphide)	0.5497
-Cl (aromatic attachment)	0.6445	-S- (one aromatic attachment)	0.0535
-Br (aliphatic attachment)	0.3997	-S- (two aromatic attachments)	0.5335
-Br (aromatic attachment)	0.8900	-SP (phosphorus attachment)	0.6270
-I (aliphatic attachment)	0.8146	-S- (two nitrogen attachments)	1.2000
-I (aromatic attachment	1.1672	-SC= (aliphatic C=)	−0.1000
Silicon		-S- (aliphatic attachment)	−0.4045
-Si- (aromatic or oxygen attachment)	0.6800	=S	0.0
-Si- (aliphatic attachment)	0.3004		

on the end of it ($f_i = −0.5473$) and an alcohol on the ring ($f_i = −0.4802$). We also need to correct for interaction between these two ortho substituents ($n_j = 1.2556$). Substituting these into the equation, we get $\log K_{ow} = 6 \times 0.294 − 0.7121 + 0.5473 − 0.4802 + 1.2556 + 0.229 = 2.603$. This is very close to the experimental value of 2.62.

In general, this approach works reasonably well for organic chemicals, but obviously not every possible structure is present in these tables, and the presence of metals in chemical compounds negatively impacts its accuracy.

'Structural contribution methods' (like this K_{ow} method), which cut molecules into pieces to estimate some property of the whole molecule, have been generated for each of the parameters shown in Table 4.1 and many others that are useful to the environmental modeller. For example, Atkinson and colleagues developed a handy method that uses structural contributions to estimate the rate at which various organic compounds react with hydroxyl radicals in the atmosphere – the principal route by which many such compounds are eliminated (Kwok & Atkinson, 1995). Some of the methods depend on each other, as shown in Figure 4.5.

4.3.3 Intra-Media Transport

Intra-media transport refers to the transport of a chemical within a given environmental compartment, like a lake, the air inside a room or the soil in a valley.

Table 4.3 Correction Factors (c_j) for Aromatic Ring Substituent Positions

Ortho Interactions	c_j	Can be either ortho or non-ortho	c_j
-COOH/-OH	1.1930	-NO$_2$ with -OH, -N<, or -N=N-	0.5770
-OH/ester	1.2556	-C≡N with -OH or -N (e.g. cyanophenols or amines)	0.5504
Amino (at 2-position) on pyridine	0.6421	-NO$_2$/-NC(=O) (cyclic-type)	0.3994
Alkyloxy (or alkylthio) ortho to one aromatic nitrogen	0.4549	-NO$_2$/-NC(=O) (non-cyclic-type)	0.7181
Alkyloxy ortho to two aromatic nitrogens (or pyrazine)	0.8955	**Non-ortho reactions**	
Alkylthio ortho to two aromatic nitrogens (or pyrazine)	0.5415	-N</-OH (e.g. 4-aminophenol)	−0.3510
Carboxamide (-C(=O)N) ortho to an aromatic nitrogen	0.6427	-N</ester (e.g. 4-aminobenzoic acid methyl ester)	0.3953
Any */-NHC(=O)C (e.g. 2-methylacetanilide)	−0.5634	-OH/ester	0.6487
Any* two/-NHC(=O)C (e.g. 2,6-dimethylacetanilide)	−1.1239	**Others**	
Any*/-C(=O)NH (e.g. 2-methylbenzamide)	−0.7352	Amino-type (at 2-position) on triazine; pyrimidine, or pyrazine	0.8566
Any* two/-C(=O)NH (e.g. 2,6-dimethylbenzamide)	−1.1284	NC(=O)NS on triazine or pyrimidine (2-position)	−0.7500
Amino-type /-C(=O)N	0.6194	1,2,3-trialkyloxy	−0.7317

* Any refers to any ortho substituent other than hydrogen (with the exception of –OH or an amino-type).

Fundamentally, there are three main transport mechanisms of relevance. At the microscopic scale, *diffusion* is the movement of a chemical from a zone of high concentration to a zone of low concentration through the influence of molecular collisions (i.e. Brownian motion). Likewise, *dispersion* is also driven by concentration gradients, but it is caused by non-ideal flow patterns (for example eddies and turbulence) and is a macroscopic phenomenon. Dispersion due to wind is typically more rapid than diffusion as a process to spread atmospheric contaminants. The terms *convection* and *advection* mean different things to different people, and are sometimes used interchangeably. In this book, *convection* is the non-random, vertical movement of a substance in the environment, for example rising warm air, or descending cold seawater. *Advection* is the lateral, non-random movement of a substance, for example wind or a flowing river. By virtue of their presence (dissolved or suspended) in a medium that happens to be flowing, chemical contaminants are often transported by natural convection and advection processes.

Intra-media transport is most rapid in water and air. In the air, convectional air currents may cause the transport of pollutant gases like CFCs into the upper

Table 4.4 Additional (Non-Aromatic) Correction Factors (c_j)

Various carbonyl factors	c_j	Various ring factors	c_j
More than one aliphatic -C(=O)OH	0.5865	1,2,3-Triazole ring	0.7525
HO-CC(=O)CO-	1.7838	Pyridine ring (nonfused)	−0.1621
-C(=O)-C-C(=O)N	0.9739	Sym-Triazine ring	0.8856
-C(=O)NC(=O)NC(=O)- (e.g. barbiturates)	1.0254	Fused aliphatic ring correction	−0.3421
-NC(=O)NC(=O)- (e.g. uracils)	0.6074	**Alcohol, ether & nitrogen factors**	
Cyclic ester (non-olefin type)	−1.0577	More than one aliphatic -OH	0.4064
Cyclic ester (olefin-type)	−0.2969	-NC(C-OH)C-OH	0.6365
Amino acid (α-carbon type)	−2.0238	-NCOC	0.5494
Di-N urea/acetamide aromatic substituent	−0.7203	HO-CHCOCH-OH	1.0649
C(C(=O)OH) aromatic (e.g. phenylacetic acid)	−0.3662	HO-CHC(OH)CH-OH	0.5944
di-N-aliphatic substituted carbamate	0.1984	-NH-NH- structure	1.1330
NC(=O)CR (R is one halogen)	0.3263	>N-N< structure	0.7306
-NC(=O)CR2+ (R is two or more halogens)	0.6365		
CC(=O)NCC(=O)OH	0.4193		
CC(=O)NC(C(=O)OH)S-	1.5505		
(Ar-O or -C-O)-CC(=O)NH-	0.4874		
>C=NOC(=O)	−1.0000		

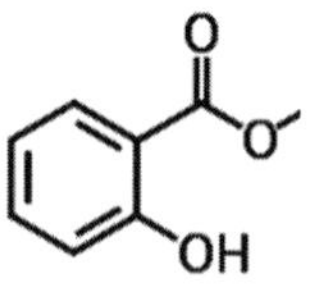

Figure 4.4   Structure of methyl salicylate

atmosphere. These chemicals then get distributed globally, so emissions from China, the EU and the United States cause the majority of the Antarctic 'ozone hole' that breaks up annually, drifts north and exposes Australians to excess UV radiation (discussed further in Section 3.5). In aquatic environments like rivers and oceans, a current may contribute to the transport of a chemical within the same compartment. For example, intra-media transport enables the contamination of sediments in the Dutch delta (Rhine, Meuse, Scheldt) by heavy metals and organic pollutants (e.g. PCBs, pesticides, PAHs) emitted upstream in Germany, Switzerland, France and Luxembourg. Note that the actual transport of the chemicals from water to sediment is an *inter*-media transport process.

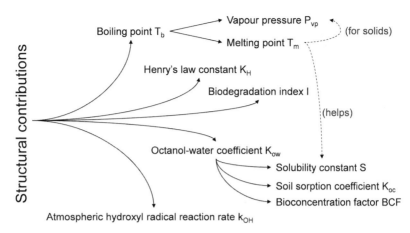

Figure 4.5 Informational dependencies between some different structural contribution methods (dotted lines – a connection is helpful for some compounds)

4.4 Transformation Processes

Transformation processes are processes where chemicals are degraded through biotic or abiotic processes. Both kinds of process have historically been modelled using simple reaction kinetics.

4.4.1 Biotic Transformation

Biotic transformation or *biotransformation* refers to the transformation of substances by or within organisms. If the organisms are microorganisms, biotransformation is usually referred to as *microbial biotransformation*. In the context of microbial biotransformation, it is possible to distinguish aerobic biodegradation (oxidative processes) and anaerobic biodegradation (reductive processes).

One example of biotransformation coupled with biological transport processes is bioturbation – in particular the mixing of sediments by animals. A wide variety of animals live in marine and freshwater sediments, sometimes at very high population densities. Some of the animals consume sediment to get at tiny organisms, but just by moving around within the sediments in their search for shelter, food and mating opportunities, sediment-dwelling animals physically blend the sediments, which can help bury contaminants, bring buried contaminants to the surface and potentially change the pH and oxygen concentrations to which the contaminants are exposed. These processes can change contaminant bioavailability. The elements selenium and arsenic are interesting examples of contaminants in sedimentary environments. As shown in Figure 4.6, after deposition or diffusion into sediment, selenium can be biotically transformed

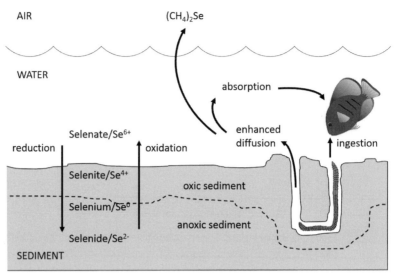

Figure 4.6 Bioturbation (e.g. sediment mixing and water pumping by sediment fauna in their burrows) releases selenium pollution by oxidising sediments and enhancing the solubility of selenium. Ions in the higher oxidation states (selenate, selenite) are more soluble in water than elemental selenium or selenide are, so bioturbation enhances mobility and bioavailability

by microbial reduction from higher oxidation states like selenate (which has oxidation state Se^{6+}) and selenite (Se^{4+}) to lower oxidation states like elemental selenium ($Se°$) and selenide (Se^{-2}). Bioturbation and water pumping through sediments by worms and other sediment fauna can reverse this process by oxidising sediments and enhancing the rate of selenium loss from contaminated sediments. Selenium in sediments exposed to oxygenated water becomes more bioavailable due to redox reactions – the more oxidised selenium compounds adsorb less strongly onto other sediment constituents, so selenate is more bioavailable than selenide. The opposite is true of arsenic, so arsenate (As^{+5}) is typically a bigger problem than arsenite (As^{3+}) under reducing conditions. Thus bioturbation will influence the transport of these two elements in opposite ways.

4.4.2 Abiotic Transformation

Abiotic transformation is the transformation of substances by abiotic processes such as hydrolysis, oxidation, reduction or photolysis. *Hydrolysis* refers to the reaction of a chemical with water (H_2O), hydrogen ions (H^+) or hydroxyl ions (OH^-). *Oxidation* refers to the electron transfer from an oxidant to the chemical to be oxidised. *Reduction* refers to the electron transfer from a reductant to the chemical to be reduced. *Photolysis* refers to the reaction of a chemical caused by photons present in sunlight.

4.4.3 Reaction Kinetics

Of course, many other chemical reactions are possible. This section describes them more generically. A *first-order reaction* depends on the concentration of only one reactant. A *second-order reaction* depends on the concentrations of one second-order reactant, or two first-order reactants. *Higher-order reactions* work accordingly. Consider the following overall reaction:

$$aA + bB \rightarrow cC + eE$$

where A, B, C and E are certain molecules and the lowercase letters indicate their stoichiometric (molar) ratios in the reaction. Depending on the detail of how the reaction proceeds, the reaction rate equations may be among the following:

$$d[A]/dt = k\,[A] \text{ (first-order reaction)} \tag{4.8}$$

$$d[A]/dt = k\,[B] \text{ (first-order reaction)} \tag{4.9}$$

$$d[A]/dt = k\,[A]^2 \text{ (second-order reaction, second-order reactant)} \tag{4.10}$$

$$d[A]/dt = k\,[B]^2 \text{ (second-order reaction, second-order reactant)} \tag{4.11}$$

$$d[A]/dt = k\,[A][B] \text{ (second-order reaction, two first-order reactants)} \tag{4.12}$$

We cannot assume that the stoichiometric ratios always correspond to the order of the reaction. For example, at 200°C, in the reaction:

$$NO_2 + CO \rightarrow NO + CO_2$$

NO_2 is a second-order reactant and CO is a zero-order reactant. In other words:

$$d[NO_2]/dt = k\,[NO_2]^2[CO]^0 = k[NO_2]^2 \tag{4.13}$$

If a second-order reaction depends on the concentrations of two first-order reactants, it may be possible to simplify from a second-order reaction to a first-order reaction. If B is in excess, [B] remains constant and we obtain a *pseudo-first-order approximation*:

$$d[A]/dt = -k'\,[A] \tag{4.14}$$

This might for example be the case if [B] is buffered by other reactions, or we are talking about a reaction of trace chemicals in water, where B is H_2O and obviously $[B] \gg [A]$. This equation is thus a pseudo-first-order approximation to the second-order reaction:

$$d[A]/dt = k\,[A][B] \tag{4.15}$$

with $k' = k\,[B]$. When such an approximation is feasible, this makes modelling easier.

4.5 Modelling Environmental Fate and Transport of Chemicals

Essentially, all models are wrong, but some are useful.

(George Box, 1987)

4.5.1 The Nature of Modelling

Models are conceptualisations of the real world, that is, they are not the real world. Prior to building a model, the modeller needs to decide which aspects of reality are relevant and which aspects of reality are to be left unaccounted for. One principle in modelling is Occam's Razor. William of Occam was an English monk who advocated simplicity both as a way of life and as a basis for logical thought. Subsequent philosophers ascribed the phrase *Entia non sunt multiplicanda sine necessitate* to him, meaning *Among competing hypotheses, the one with the fewest assumptions should be selected.* For environmental scientists, using Occam's Razor means avoiding unnecessary complexity and excluding environmental processes that do not clearly play a major role. Model quality can be undermined by lack of data and uncertainty resulting from complexity.

4.5.2 Describing Environmental Models

Two broad groups of models are commonly used to address environmental fate and transport of chemicals: compartment models and plume models.

4.5.2.1 Compartment Models

Compartment models, also referred to as Box models or Mackay models, conceptualise the environment as consisting of homogeneous, well-mixed compartments. The principle of mass conservation and mass balance equations are the basis for compartment models. Once a chemical is emitted into a given compartment, it can be transferred to other compartments (Figure 4.7). The USEtox™ model (www.usetox.org) is designed to characterise human and ecotoxicological impacts in life cycle assessment and comparative risk assessment. USEtox is an example of a compartment model which, in the

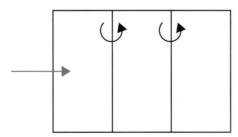

Figure 4.7   Conceptual representation of a model with three compartments. Chemicals enter a given compartment (straight arrow) and can be transferred into other compartments (curled arrows). As soon as a chemical enters a given compartment, the concentration of the chemical in that compartment is assumed to immediately be homogeneously distributed throughout the whole compartment

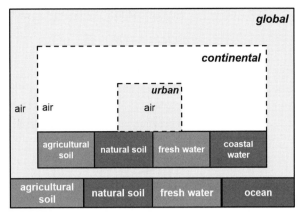

Figure 4.8 Environmental compartments featured in the USEtox™ model (illustration adapted from Fantke et al., 2017)

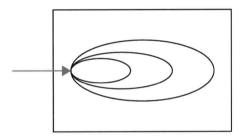

Figure 4.9 Two-dimensional representation of a pollutant plume within a rectangular compartment

original version shown in Figure 4.8, features 11 compartments on two geographical scales.

4.5.2.2 Plume Models

Plume models are used if it is assumed that compartments are not homogeneous and not well mixed. Plume models aim to model the distribution of a chemical within a given compartment. Figure 4.9 is a simple representation of the emission of a contaminant from the left to the right side of the figure, and its spreading due to diffusion and dispersion effects. One could imagine this happening in a lake with a current from the west, or in groundwater flowing downhill to the east, or in the air above a city with a westerly breeze.

Of course scientists have developed much more complex models of pollutant plumes by including the third and fourth (time) dimensions. Figure 4.10 represents a buoyant plume that continues rising in the atmosphere for some time in the z-dimension after being emitted by a chimney, spreading out in the y-dimension and being dispersed by a wind blowing in the x-dimension.

A continuous, buoyant air pollution plume like this can be modelled using Equation 4.16.

$$C = \frac{Qf}{u\sigma_y\sqrt{2\pi}} \frac{g_1 + g_2 + g_3}{\sigma_z\sqrt{2\pi}} \tag{4.16}$$

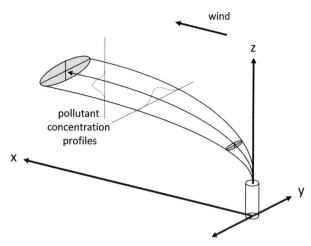

Figure 4.10 Dispersion in three dimensions of an atmospheric plume emitted by a chimney

This enables environmental scientists to estimate the concentration C [g.m^{-3}] of a substance emitted at any point in space, when the substance is emitted at a continuous rate Q [g.s^{-1}] and blown by a wind with speed u [m.s^{-1}] in direction x. The horizontal and vertical dispersion coefficients (σ_y and σ_z) of the emission distribution are functions of the distance x from the source and the stability (or turbulence) of the atmosphere. The vertical dispersion parameters g_i depend on these coefficients, the height of the chimney and the height of any inversion layer in the atmosphere.

But it gets much more complicated than this. Just some of the important computational models for atmospheric dispersion which have been accepted by national environmental regulatory agencies include AERMOD and CALPUFF, which were developed in the United States, Germany's ATSTEP and Australia's AUSPLUME, but there are hundreds more. The interested reader can learn more about atmospheric dispersion modelling from books on environmental physics such as Boeker and von Grondelle (1995), or books dedicated to air pollution such as those of Beychok (2005) or Turner (1994).

4.5.2.3 Steady State and Unsteady State

The first thing an analyst should identify when trying to model pollution in an environmental system is whether the system is at *steady state*. In such cases, the concentration of the pollutant is constant despite ongoing processes that strive to change the concentration. When for instance a chemical is continuously discharged into a previously pristine, freshwater lake, the concentration of the chemical increases until, at some point in time, the steady state concentration is reached. This means that the mass of chemical added equals the mass of chemical removed through processes such as evaporation, run-off, sedimentation or degradation. When the emission of a chemical to a lake is reduced, the concentration of the chemical decreases until, at some point in time, steady state is reached again.

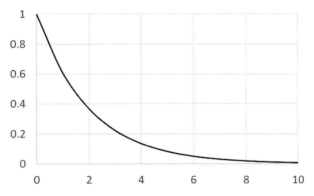

The down-going curve

In mathematical terms, steady state is described by:

$$F_{in} = F_{out} \tag{4.17}$$

$$dM/dt = 0 \tag{4.18}$$

where F_i is a flowrate [mass.time^{-1}], M is the mass of contaminant in the lake and t is time. Unsteady state is described by:

$$F_{in} \neq F_{out} \tag{4.19}$$

$$dM/dt = (F_{in} - F_{out}) \tag{4.20}$$

4.5.2.4 The Down-Going Curve

If the emission of a chemical to a given compartment ceases while the chemical is still being removed from the compartment through processes such as evaporation, sedimentation or degradation, it is possible to calculate the concentration of the chemical in the given compartment after a given period of time. This is a simple unsteady state system. The shape of the curve is shown in Figure 4.11 and follows from the following mathematical derivation:

$$F_{in} = 0 \tag{4.21}$$

$$F_{out} = kM \tag{4.22}$$

$$dM/dt = F_{in} - F_{out} = - F_{out} = - kM \tag{4.23}$$

Dividing by volume, this becomes:

$$dC/dt = - F_{out}/V = - kC \tag{4.24}$$

You can see this is the same format as the first-order rate Equation 4.8, since it is a first-order rate process. If we rearrange the equation, we can get:

$$dC/C = -k\, dt \tag{4.25}$$

Integrating this, we get:

$$\ln(C) - \ln(C_0) = -kt \tag{4.26}$$

and if you take the exponential of that, we conveniently get:

$$C = C_0 \, e^{-kt} \tag{4.27}$$

Of course, if there are several loss processes (in this example: k_i where $i = 1$ for evaporation, $i = 2$ for sedimentation, and $i = 3$ for degradation) that depend on the amount of the chemical in the lake, then the initial mass balance equation is:

$$dM/dt = F_{in} - F_{out} = F_{in} - k_1 \, M - k_2 \, M - k_3 \, M \tag{4.28}$$

It follows that:

$$C = C_0 e^{-(k_1 + k_2 + k_3)t} \tag{4.29}$$

One of the other common loss processes we can describe in this way is when there is a flow through the compartment, for example if the compartment is a lake with inflows and precipitation entering it and a creek running out of it. Since we have assumed the contaminant has ceased to flow into the lake, a useful point to recognise in cases like this is that the first-order rate constant k [s^{-1}] for this loss process is the inverse of the residence time τ [s]:

$$\frac{1}{k} = \tau = \frac{compartment\ volume\ [m^3]}{volumetric\ flowrate\,[m^3.s^{-1}]} \tag{4.30}$$

4.5.2.5 The Up-Going Curve

If a chemical is emitted to a compartment which has previously been uncontaminated by the given chemical, the concentration of the chemical starts at zero, and if there are first-order loss processes such as those described previously, eventually the concentration approaches a steady state. The shape of the up-going curve is shown in Figure 4.12 and follows from the following mathematical derivation:

$$F_{in} > 0 \tag{4.31}$$

$$F_{out} = kM \tag{4.32}$$

$$dM/dt = F_{in} - F_{out} = F_{in} - kM \tag{4.33}$$

$$dC/dt = F_{in}/V - kC \tag{4.34}$$

Following the same process as for the down-going curve, we get:

$$C = C_{max} \left[1 - e^{-kt} \right] \tag{4.35}$$

4.5.2.6 Numerical Approaches

While simple cases can be solved analytically when the boundary conditions fit with the conditions underlying the down-going or up-going curves, an analytical solution

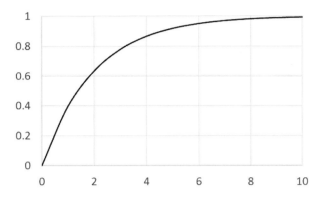

Figure 4.12   The up-going curve

may not be tractable for more complicated cases. Say a lake has a certain concentration, and the release of a chemical is intermittent and gradually increasing over time. Here, numerical solutions may be more straightforward. With numerical solutions, the concentration of the chemical is calculated in an iterative fashion:

$$C(t+1) = f[C(t)] \qquad (4.36)$$

In other words, the concentration after a short time has elapsed is a function of what the concentration was a moment ago. In some of the examples in the next section, numerical approaches are demonstrated and contrasted with analytical approaches.

4.6 Mathematical Examples

4.6.1 Intra-Media (Phase) Transfers

Given values for constants like K_{ow}, H and K_s that describe some equilibrium condition, and data on a system that is currently not in equilibrium, it is possible to do calculations to determine the mass of a chemical that will be transported across some phase boundary. For example, imagine you are thirsting for the water in a well that is contaminated with metaldehyde (a pesticide). The water has a metaldehyde concentration of 0.1 mg/L, but if you add activated carbon to adsorb the contaminant and then filter out the activated carbon, you can reduce the dissolved contaminant concentration. If you want to drink a glass of this water, how much activated carbon would you need to add to it to reduce the concentration to a safe level (say, 0.001 mg/L)?

To answer this question, we recognise that it implies adsorbing the equivalent of 0.099 mg/L to the activated carbon (that is, 0.02475 mg of metaldehyde in a 250 mL glass). If we know from the scientific literature that $K = 0.4$ (mg/g)(L/mg)$^{1/n}$ and $n = 3.3$, then Freundlich says:

$$x = mKC^{1/n} \qquad (4.37)$$

$$0.02475 = m \times 0.4 \times 0.001^{1/3.3} \quad \text{and} \qquad (4.38)$$

$$m = 0.502 \text{ g} \qquad (4.39)$$

So less than a teaspoon should do it. With a bit more data, constants like these can also help us to calculate the rate at which certain environmental processes will occur. Instead of the well, imagine it is a lake that is contaminated by metaldehyde. The evaporative flux of the contaminant from the surface of the lake into the atmosphere is driven by the concentration difference between the air and water, and constrained by diffusional effects at the phase boundary that determine the velocity of mass transfer. For water-to-air transfers, we assume the mass transfer rate is controlled by the denser medium (the water). According to Liu et al. (2015):

$$R = S \times K_{OL}\left(C_{water} - \frac{C_{air}}{H'}\right) \qquad (4.40)$$

where R is the emission rate (mg/hr), S is the surface area (m^2) and K_{OL} is the overall liquid-phase mass transfer velocity (m/hr). Obviously, if the system is at equilibrium, $H'C_{water} = C_{air}$ (Henry's law – notice we are using a dimensionless expression here, inverted compared to Section 4.3.2) and no mass transfer

$$\frac{1}{K_{OL}} = \frac{1}{k_L} + \frac{1}{k_G H'} \qquad (4.41)$$

where k_L and k_G are the liquid-phase and gas-phase mass transfer velocities (m/hr). Hites (2007) provides some formulae for the estimation of these:

$$k_L = 0.0004(0.1u^2 + 1)\left(\frac{32}{mw}\right)^{0.5} \qquad (4.42)$$

$$k_G = (0.2u + 0.3)\left(\frac{18}{mw}\right)^{0.5} \qquad (4.43)$$

with units of cm/s, where u (m/s) is the airspeed and mw (g/mol) is the molecular weight of the contaminant. Imagine a small lake of 1,000 m^2, contaminated like the well in the previous example. It is in a tropical country, the air and water are at 25°C (298 Kelvin) and the wind speed is the global average used in the USEtox model: 3 m/s. Using that value and the molecular weight of the contaminant (176 g/mol), we can calculate that k_L and k_G are 3.24 x 10^{-4} and 0.288 cm/s respectively. Then if H' is 2.13 x 10^{-3}:

$$\frac{1}{K_{OL}} = \frac{1}{3.24 \times 10^{-4}} + \frac{1}{0.288 \times 2.13 \times 10^{-3}} = 4718 \text{ s/cm} \qquad (4.44)$$

so after inversion, K_{OL} is 2.12 x 10^{-4} cm/s. Coming back to the emission rate equation. We assume a low concentration of metaldehyde in the air downwind from a farm (10^{-5} mg/L) which has little impact on the emission rate.

$$R = 1000[m^2] \times 2.12 \times 10^{-4}\left[\frac{cm}{s}\right] \cdot \frac{1}{100}\left[\frac{m}{cm}\right]\left(0.1 - \frac{10^{-5}}{2.13 \times 10^{-3}}\right)\left[\frac{mg}{L}\right] \cdot 1000\left[\frac{L}{m^3}\right]$$

$$= 0.21 \text{ mg/s} \qquad (4.45)$$

4.6.2 Lake Pollution: Inter-Compartment Flows

Let us continue thinking about dissolved pollution in water bodies. Lakes may receive various pollutants through a range of different pathways (e.g. atmospheric deposition, inflowing streams, point discharges). Say a new factory is to be built and will discharge a certain amount of a certain chemical to the lake. One may for example be interested in the concentration of the pollutant in the lake water that will eventually be reached given the emission of the pollutant by the new factory. This concentration would be of interest if the lake served as a drinking water source. In this example, the following numbers are provided:

Volume of the lake:	V_L	$=$	$1 \cdot 10^6$	m^3
Surface area of the lake:	A_L	$=$	$6 \cdot 10^4$	m^2
Water inflow to the lake:	Q_I	$=$	$5 \cdot 10^{-3}$	$m^3\,s^{-1}$
Evaporation from the lake:	v_E	$=$	1	$m\,yr^{-1}$
Planned pollutant discharge rate:	m_F	$=$	40	$t\,yr^{-1}$

It is furthermore assumed that the lake water stock is in steady state, that is, the difference between inflow and evaporation will flow out of the lake in an outlet stream or underground seepage. We can calculate the evaporation rate from the lake:

$$Q_E = A_L \cdot v_E = 6 \times 10^4 \, [m^2] \times 1 \left[\frac{m}{yr} \right] \times \frac{1}{365} \left[\frac{yr}{day} \right] \times \frac{1}{86400} \left[\frac{day}{s} \right]$$
$$= 1.9 \times 10^{-3} \left[\frac{m^3}{s} \right] \tag{4.46}$$

We can also calculate the rate Q_O at which water leaves the lake through the outlet stream and underground seepage:

$$Q_O = Q_I - Q_E = (5 - 1.9) \times 10^{-3} \left[\frac{m^3}{s} \right] = 3.1 \times 10^{-3} \left[\frac{m^3}{s} \right] \tag{4.47}$$

4.6.2.1 Case A: Pristine Lake

We first consider a situation where the lake is pristine and there are no other sources of the pollutant besides the factory discharge. Also, we know that the pollutant under consideration does not evaporate, that is, the pollutant concentration in the evaporating water is zero. As long as we do not wish to make any statement about how long it takes until a certain pollution level is reached, we can base our calculation on the mass-balance principle under eventual steady state conditions ($F_{in} = F_{out}$). Note that in this example, there is no chemical present in the inflow ($C_I = 0$), and the pollutant does not evaporate ($C_E = 0$):

$$Q_I \times C_I + m_F = Q_E \times C_E + Q_O \times C_O \Rightarrow m_F = Q_O \times C_O \Rightarrow C_O = \frac{m_F}{Q_O} \tag{4.48}$$

If we use the numbers provided earlier, we obtain $C_L = C_O = 409 \text{ g/m}^3$.

Alternatively, we may want to know the concentration of the pollutant in the lake water after the residence time of water in the lake. The residence time of the lake water stock can be calculated as follows:

$$\tau_L = \frac{V_L}{Q_I} = \frac{1 \times 10^6 \, [m^3]}{5 \times 10^{-3} \, [\frac{m^3}{s}] \, 86400 \, [\frac{s}{d}]} = 2315 \text{ days} \qquad (4.49)$$

In order to calculate the concentration of the pollutant in the lake water after the residence time of the lake water stock, we can no longer assume steady state conditions. Rather, we must resort to a numerical approach or the equation for the up-going curve.

Numerical Approach

We first consider a numerical approach. In the current example, the initial concentration of the pollutant in the lake water is zero:

$$C_L(t=0) = 0 \, [g/m^3] \qquad (4.50)$$

For the numerical calculation, we assume that the pollutant is discharged at once at the start of every day and then is immediately distributed homogeneously in the whole lake water stock. The daily pollutant emission is as follows:

$$\Delta m_m = m_F \times 1[day] = 40 \left[\frac{tonne}{year}\right] \times 10^6 \left[\frac{g}{tonne}\right] \times \frac{1}{365} \left[\frac{year}{day}\right] \times 1[day] = 109589g \qquad (4.51)$$

After the pollutant is emitted, the concentration in the lake is as follows:

$$C_L(t=0^+) = C_L(t=0) + \frac{\Delta m_{in}}{V_L} = 0 \left[\frac{g}{m^3}\right] + \frac{109589[g]}{1 \times 10^6 [m^3]} = 0.11 \text{ g/m}^3 \qquad (4.52)$$

The removal of the pollutant by the outlet stream or underground seepage depends on the pollutant concentration in the lake. Although the pollutant is removed from the lake continuously during the whole day through the outlet stream or underground seepage, we approximate this situation by assuming that the pollutant is removed all at once at the end of the day:

$$C_L(t=1^-) = C_L(t=0^+) \qquad (4.53)$$

where 1⁻ means "just before 1" and 0⁺ means "just after 0". Then

$$\Delta m_{out}(t=1^-) = Q_O \times C_L(t=1^-)$$
$$= 3.1 \times 10^{-3} \left[\frac{m^3}{s}\right] \times 86400 \frac{[s]}{[day]} \times 1[day] \times 0.11 \frac{[g]}{[m^3]} = 29.46 \text{ g} \qquad (4.54)$$

At the end of the given day, the concentration in the lake is as follows:

$$C_L(t=1) = C_L(t=1^-) - \frac{\Delta m_{out}(t=1^-)}{V_L} = 0.110 \frac{[g]}{[m^3]}$$
$$-2.95 \times 10^{-5} \left[\frac{g}{m^3}\right] = 0.110 \text{ g/m}^3 \qquad (4.55)$$

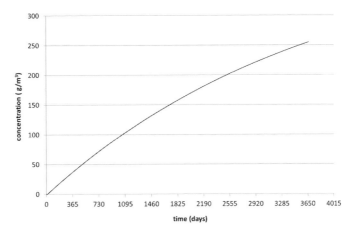

Figure 4.13   Pristine lake contamination example

This procedure can be iterated until t = 2315 d (see Figure 4.13). The following lake water concentration is found after the residence time of the lake water stock:

$$C_L(t = 2315) = 189 \text{ g/m}^3 \tag{4.56}$$

Analytical Approach

The same result can be obtained much faster by using the equation for the up-going curve:

$$C = C_{max}(1\text{-e}^{-kt}) \tag{4.57}$$

The first-order removal rate constant k is the inverse of the residence time of the chemical:

$$k = \frac{1}{\tau_C} \tag{4.58}$$

$$\tau_C = \frac{V_L}{Q_O} = \frac{1 \times 10^6 [m^3]}{3.1 \times 10^{-3} \left[\frac{m^3}{s}\right] \times 86400 \left[\frac{s}{day}\right]} = 3734 \text{ days} \rightarrow k = 2.68 \times 10^{-4} \text{day}^{-1}$$

$$\tag{4.59}$$

The steady state concentration C_{max} is a function of the residence time of the chemical, the rate at which the pollutant is added and the volume of the lake:

$$C_{max} = \frac{\tau_C \times m_F}{V_L} = \frac{3734[days] \times 40\left[\frac{tonne}{year}\right] \times 10^6\left[\frac{g}{tonne}\right]}{365\left[\frac{days}{year}\right] \times 10^6[m^3]} = 409 \text{ g/m}^3 \tag{4.60}$$

The solution to the up-going equation thus is:

$$C_L(t = \tau_L) = C_{max}(1 - e^{-k\tau_L}) \tag{4.61}$$

$$C_L(t = 2315) = 409 \left[\frac{g}{m^3}\right] (1 - e^{-2.68 \times 10^{-4}[day^{-1}] \times 2315[day]}) = 189 \text{ g/m}^3 \quad (4.62)$$

4.6.2.2 Case B: Previous Diffuse Pollution

Let us now consider an alternative situation where the pollutant of concern is also present through other sources (diffuse pollution) before the new factory starts operation.

Rate of diffuse pollution: $m_D = 10$ tonne/year

Before the new factory starts discharging more of the pollutant, the lake is in steady state. The steady state concentration is:

$$C_{L,pre} = \frac{10 \left[\frac{tonne}{year}\right] \times 10^6 \left[\frac{g}{tonne}\right] \times \frac{1}{365} \left[\frac{year}{day}\right] \times \frac{1}{86400} \left[\frac{day}{s}\right]}{3.1 \times 10^{-3} \left[\frac{m^3}{s}\right]} = \frac{0.317 \left[\frac{g}{s}\right]}{3.1 \times 10^{-3} \left[\frac{m^3}{s}\right]}$$

$$= 102 \text{ g/m}^3 \quad (4.63)$$

The new steady state concentration in the lake is of course higher and could be calculated based on a mass-balance equation with additional terms added:

$$C_{L,B} = \frac{(10 + 40) \left[\frac{tonne}{year}\right] \times 10^6 \left[\frac{g}{tonne}\right] \times \frac{1}{365} \left[\frac{year}{day}\right] \times \frac{1}{86400} \left[\frac{day}{s}\right]}{3.1 \times 10^{-3} \left[\frac{m^3}{s}\right]} = \frac{1.585 \left[\frac{g}{s}\right]}{3.1 \times 10^{-3} \left[\frac{m^3}{s}\right]}$$

$$= 511 \text{ g/m}^3 \quad (4.64)$$

If we apply the same mass-balance equation as in case A, we would merely obtain the concentration change between the old steady state and the new steady state:

$$\Delta C_L = 409 \text{ g/m}^3 \quad (4.65)$$

This could of course be combined with the steady state concentration before the factory started in order to obtain the new steady state concentration after the factory started:

$$C_{L,B} = C_{L,pre} + \Delta C_L = 102 \left[\frac{g}{m^3}\right] + 409 \left[\frac{g}{m^3}\right] = 511 \text{ g/m}^3 \quad (4.66)$$

We may still want to know the concentration of the pollutant in the lake water after the residence time of the lake water stock, or rather, how much the concentration will increase due to the additional discharges from the new factory. Again, we can use a numerical approach or the up-going curve.

The total load after the factory started operation is:

$$m = m_D + m_F = 10 \left[\frac{tonne}{year}\right] + 40 \left[\frac{tonne}{year}\right] = 50 \text{ tonne/year} \quad (4.67)$$

Numerical Approach

Let us first consider the numerical approach and change the boundary conditions:

$$C_{O,B}(t=0) = 102 \text{ g/m}^3 \qquad (4.68)$$

$$\Delta m_{in} = m \times 1[day] = 50\left[\frac{tonne}{year}\right] \times 10^6\left[\frac{g}{tonne}\right] \times \frac{1}{365}\left[\frac{year}{day}\right] \times 1[day]$$
$$= 136986 \text{ g} \qquad (4.69)$$

Mimicking the process shown in equations 4.52-4.55, the following lake water concentration is found after the residence time of the lake water stock:

$$C_{L,B}(t=2{,}315) = 291 \text{ g/m}^3 \qquad (4.70)$$

Note that one could basically also use the calculations from case A and simply add the steady state concentration prior to the start of the factory operation to every single time step:

$$C_{L,B}(t) = C_{L,A}(t) + C_{L,pre} = C_{L,A}(t) + 102\left[\frac{g}{m^3}\right] \qquad (4.71)$$

The concentration after the residence time of the lake water stock and the concentration increase due to the new factory would thus be:

$$C_{L,B}(t=2315) = C_{L,A}(t=2315) + C_{L,pre} = 189\left[\frac{g}{m^3}\right] + 102\left[\frac{g}{m^3}\right] = 291\left[\frac{g}{m^3}\right] \qquad (4.72)$$

Analytical Approach

When directly using the up-going curve, we must realise that we cannot simply use the up-going curve as if the concentration were zero at time zero. Rather, we need to figure out what time on the up-going equation for the new steady state corresponds with the old steady state:

$$C_{L,pre} = C_{max,B}(1 - e^{-kt^*}) \Rightarrow t^* = -\ln\left[1 - \frac{C_{L,pre}}{C_{max,B}}\right] \times \frac{1}{k} \qquad (4.73)$$

$$t^* = -\ln\left[1 - \frac{102\left[\frac{g}{m^3}\right]}{511\left[\frac{g}{m^3}\right]}\right] \times \frac{1}{2.86 \times 10^{-4}[d^{-1}]} = 831\text{d} \qquad (4.74)$$

Now we can use the equation for the up-going curve once again to estimate the concentration 2,315 days after the factory started production:

$$C'_{L,B}(t=2315+t^*) = C_{max,B}\left(1 - e^{-k(2315+t^*)}\right) \qquad (4.75)$$

$$C'_{L,B}(t=2315+t^*) = 511\left[\frac{g}{m^3}\right]\left(1 - e^{-2.68\times10^{-4}[day^{-1}](2315+831)[day]}\right) \qquad (4.76)$$

$$C_{L,B}(t = 2315) = C'_{L,B}(t = 2315 + t^*) = 291 \text{ g/m}^3 \qquad (4.77)$$

The same values can be obtained by using the up-going curve from case A. Here too, one simply adds the steady state concentration from before the factory started operation:

$$C_{L,B}(t = 2315) = C_{L,A}(t = 2315) + C_{L,pre} = 189\left[\frac{g}{m^3}\right] + 102\left[\frac{g}{m^3}\right] = 291 \text{ g/m}^3$$

$$(4.78)$$

4.6.3 Soil Pollution: Phase Transfer, Reaction and Flow

Under current conventional agricultural practices, pesticides are routinely applied to agricultural fields and crops in order to prevent disease. Say a farmer applies pesticide to a field once every four weeks. The pesticide evaporates, is microbially degraded and leaches to the groundwater body:

Pesticide application (every four weeks):	F_P	=	1	kg/ha
Pesticide evaporation rate:	k_E	=	0.01	d^{-1}
Pesticide evaporation rate:	k_D	=	0.03	d^{-1}
Leaching to groundwater:	k_L	=	0.02	d^{-1}

We may now be interested in how the pesticide concentration develops over time. We assume a homogeneous soil compartment with a certain thickness. The thickness of the soil compartment is commonly chosen as: $d_S = 20$ cm.

The calculation is possible by using a numerical approach, or by using the differential equation for the down-going curve to account for losses after each application. Note, though, that given the intermittent pesticide application, we need to consider the curve segment associated with each four-week period separately:

$$C_S(t) = C_{S_i} \times e^{-k(t-t_i)} \text{ with } k = k_E + k_L + k_D = 0.06 \text{ day}^{-1} \qquad (4.79)$$

The pesticide concentration in the soil after the first application ($t_1 = 0[\text{d}]$) is as follows:

$$C_S(t_1^+) = \frac{F_P}{d_S} = \frac{1\left[\frac{kg}{ha}\right] \times \frac{1}{10000}\left[\frac{ha}{m^3}\right] \times 1000\left[\frac{g}{kg}\right]}{20[cm]\frac{1}{100}\left[\frac{m}{cm}\right]} = 0.5 \text{ g/m}^3 \qquad (4.80)$$

During the following four-week period, that is between $t_1 = 0$ days and $t_2 = 28$ days, the pesticide concentration evolves as follows:

$$C_S(t) = C_{S_1} \times e^{-k(t-t_1)} \qquad (4.81)$$

The concentration in the soil at $t_2 = 28$ days, just before application of the next batch of pesticide, is as follows:

$$C_S(t_2^-) = 0.500 \times e^{-0.06 \times (28-0)} = 0.093 \text{ g/m}^3 \qquad (4.82)$$

The concentration in the soil at $t_2 = 28$ days, just after the second pesticide application, is as follows:

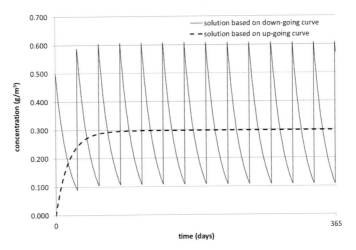

concentration (g/m³)

time (days)

— solution based on down-going curve
- - solution based on up-going curve

Figure 4.14 Pesticide concentration in soil: Example

$$C_S(t_2^+) = C_S(t_2^-) + \frac{F_P}{d_S} = 0.593 \text{ g/m}^3 \qquad (4.83)$$

During the following four-week period, that is between $t_2 = 28$ days and $t_3 = 56$ days, the pesticide concentration evolves as follows:

$$C_S(t) = C_{S_2} \times e^{-k(t-t_2)} \qquad (4.84)$$

The concentration in the soil at $t_3 = 56$ days, just before application of the next batch of pesticide, is as follows:

$$C_S(t_3^-) = 0.593 \times e^{-0.11(56-28)} = 0.11 \text{ g/m}^3 \qquad (4.85)$$

The concentration in the soil at $t_3 = 56$ days, just after the third pesticide application, is as follows:

$$C_S(t_3^+) = C_S(t_3^-) + \frac{F_P}{d_S} = 0.610 \text{ g/m}^3 \qquad (4.86)$$

The steps detailed earlier can be continued until every subsequent period looks like the one before. This is shown as the spikey curve in Figure 4.14. The same results could be obtained using a numerical approach like the one in the previous (lake) example.

If one were simply interested in how the average soil concentration develops, this could also be calculated using the differential equation for the up-going curve, where C_{max} can be obtained through the steady state mass balance equation.

$$\frac{F_P}{d_S} \frac{1}{28[day]} = C_{max} \times k \rightarrow C_{max} = 0.298 \text{ g/m}^3 \qquad (4.87)$$

Once C_{max} is calculated, it is possible to plot the up-going curve:

$$C = C_{max}(1 - e^{-kt}) \text{ with } k = k_E + k_L + k_D = 0.06 \text{ day}^{-1} \qquad (4.88)$$

This curve is also plotted in Figure 4.14 and illustrates the approximation of the actual curve by this time-averaged curve. Note that with this last approach, no consideration is given to the fact that the concentration changes during every period of application and losses. Still, this approach gives a quick indication of how many application periods it takes for the average concentration to stabilise.

4.7 The Need for Simplicity

This chapter describes some of the key processes influencing contaminant transport and fate – processes that help contaminants move between different phases and locations in the environment. It also describes elements of simple models that are used to predict the behaviour of pollution. Models of pollution can be much more complicated than this, but sometimes simple models are best when the amount of observational data we have is limited. The alternative would involve making a lot of assumptions and disobeying the principles of thinkers from William of Occam (see Section 4.5.1) to Antoine de Saint-Exupéry. The latter wrote, 'perfection is finally attained not when there is no longer anything to add, but when there is no longer anything to take away' (Saint-Exupéry, 1939). The process of getting scientists to agree on the simplest set of assumptions for an environmental model has been called 'the search for harmony and parsimony' (Hauschild et al., 2008). It is an ongoing search.

In the meantime, although environmental models are always going to be approximations, they are necessary tools for risk assessors who need to inform the priorities of governments and companies who could spend money on a variety of different environmental protection initiatives, or who need to choose the least dangerous product for a particular purpose. Chemical risk assessment is the subject of the next chapter in this book (Chapter 5). Simple environmental models are also built into the impact assessment processes used in life cycle assessment (Chapter 7).

4.8 Review Questions

1. Describe two different models for adsorption using simple equations.
2. Explain how contaminants in sediments can be more or less bioavailable. What processes can change their bioavailability?
3. What are structural contribution methods useful for, and what kinds of parameters can we get from them?
4. Look up the structure and K_{ow} of DDT. Then try to estimate K_{ow} using the data in this chapter. How close did you get?
5. What kind of a model is USEtox? What is the other main kind? Why would you use one or the other kind?
6. If a lecture theatre has been locked up with the air-conditioning off, and is full of musty-smelling air, what kind of equation would describe the decrease of the

musty smell when the doors are opened? Given the flowrate Q [m³/s] through the door and the volume of the theatre [m³], how would you calculate the key coefficient for this process?

7. A coastal city on a plain 3 km by 3 km and ringed by mountains suffers a temperature inversion 500 m above the ground. If cars emit 2 tonnes of pollution to the air each day, and the only thing with a pollution-reducing effect is a sea breeze that exchanges 25% of the city air each day, what is the concentration of pollution in the air? (Assume that the air mixes instantly throughout the airshed.)

8. Imagine a variation on the previous question in which there is no breeze for a day and the concentration of the pollution reaches 3 mg/m³. If the breeze then comes back, what is the concentration of pollution in the air a day after the breeze returned?

5 Introduction to Toxicology

Toxicology is the study of how chemicals can affect the health of humans and the environment. This chapter assumes a basic knowledge of chemistry and chemical reactions, and focusses on what happens when an animal is exposed to a toxic chemical (or 'toxicant'). The Swiss-German philosopher Paracelsus (1493–1541) is regarded as the father of toxicology. He said: 'All substances are poisons; there is none which is not a poison. The right dose differentiates a poison and a remedy.' There are plenty of examples of this – arsenic is a convenient illustration: it has been used for centuries to poison unwanted spouses, political rivals and agricultural pests, but it has also had useful therapeutic applications in the treatment of sexually transmitted disease and cancer. So an understanding of the rate at which a person is exposed to a toxicant, the focus of the previous chapter, is important in understanding whether a toxic effect will arise. Nevertheless, if you are exposed to a particular toxicant, the kind of health effects it will have on you will depend on how it reacts with the chemistry of your cells, how it is transported around your body and the ways in which your body defends itself. Understanding these aspects of toxicology will in turn improve your ability to assess the risk the chemical poses – which is the subject of the next chapter. For the sake of simplicity and immediacy, this chapter focusses on you, the reader, as the subject of toxicological study, i.e. the animal we call *Homo sapiens sapiens*, but the material is relevant also to other living beings.

One may divide the study of toxicology into two principal components: toxicokinetics and toxicodynamics. Toxicokinetics concerns how a toxicant enters the body, how it is distributed, modified by metabolic processes and excreted. The rate at which these processes occur will determine how long the toxicant is inside you in a damaging form. Toxicodynamics is the study of how a toxicant impacts the biological function of your cells. It may do this by binding to proteins or nerve receptors, mimicking other chemicals, altering the structure of your normal biochemicals or via other means.

5.1 Toxicokinetics

5.1.1 Absorption

For a toxicant to impact your health, it must either damage you from the outside or overcome your outer defences and enter your body. As an example of the first kind of impact, concentrated sulphuric acid, like other corrosive acids, can destroy your skin

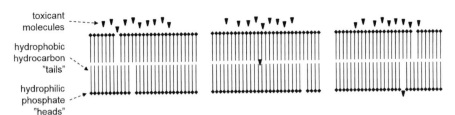

Figure 5.1 Time sequence (left to right) showing diffusion of a toxicant molecule through a biological membrane

by hydrolysis of the proteins and lipids in it. This is a *local effect*. On the other hand, when a toxicant like alcohol affects the cells of your liver, toxicologists would call this a *systemic effect* – one which occurs in a different organ to the organ that first absorbed the toxicant. The principal toxicant pathways into the human body are the lungs, the digestive tract and the skin. Each of these surfaces presents a membrane barrier that the toxicant has to pass through. So how can an unwanted molecule get through such membranes? The same ways all the necessary molecules do: diffusion, filtration, facilitated diffusion, active transport or endocytosis.

5.1.1.1 Diffusion and Filtration

Any molecule that is at a temperature above absolute zero ($-273°C$) is in motion. A water molecule at normal body temperature ($37°C$) is whizzing around at 2,500 km/h (Birgersson et al., 1995). Despite the speed, from a human perspective, it does not get far because of all the other molecules it collides with. These collisions result in the molecule taking an apparently random, zig-zag path we call *Brownian motion*. The phenomenon of Brownian motion has the consequence that molecules naturally *diffuse* from a place where they are at a high concentration to a place with a lower concentration, until there is no difference in concentration.

Suppose you are suddenly exposed to a concentrated toxicant liquid. When considered in relation to the idea of a toxicant trying to pass through a biological membrane (like the membranes in your skin) and into your body, the phenomenon of Brownian motion has two important consequences. First, there is a concentration difference between the outside of the membrane and the inside, which creates the possibility of diffusion in that direction. Secondly, the molecules that make up the membrane are also in constant motion, which may sometimes lead to tiny gaps between the molecules. This is illustrated in the triptych in Figure 5.1. This shows a typical biological membrane consisting of a double row of surface-active, phospholipid molecules, each with a hydrophilic phosphate head facing outwards and a long hydrophobic, hydrocarbon tail facing into the middle. These molecules naturally line up as shown when phospholipids are exposed to water, with the hydrophilic heads pointing towards it and the hydrophobic tails pointing away from it.

A gap appears momentarily in the outer surface of the membrane (above it) as shown on the left, and the random movements of a toxicant molecule send it into the gap. The toxicant molecule bumps around inside the membrane (middle picture) until a gap appears next to it on the inside surface of the membrane (right-hand picture).

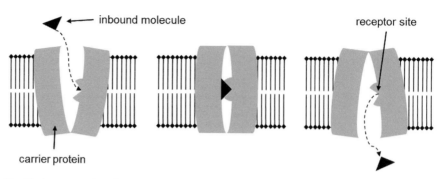

inbound molecule

receptor site

carrier protein

Figure 5.2 Simplified representation of a carrier protein in operation

The concentration of the toxicant on the inner surface of the membrane is zero, so the molecule goes through.

Since the interior of the membrane consists of hydrophobic hydrocarbons, fat-soluble toxicants are more likely to make this trip than hydrophilic molecules. Furthermore, small molecules are more likely to pass through the membrane than large ones. Carbon dioxide and oxygen are examples of small molecules (about 0.4 and 0.3 nm across, respectively) that may easily diffuse through the membrane. As it is with these gases, the rate at which a certain toxicant passes through the membrane is proportional to the concentration difference across it.

Many large molecules and hydrophilic toxicants can also pass through membranes, a fact that indicates that diffusion is not the only process at work at membrane barriers. *Active transport, facilitated diffusion, endocytosis* and *exocytosis* are other key processes in these cases, and are discussed shortly. But the simplest of all is *filtration*. In this context, *filtration* refers to transport through large pores in the membrane. Some of the pores are large enough for water to flow in and out in response to pressure changes. Small, dissolved toxicant molecules can therefore flow in or out through them without the membrane interacting with them or being selective about what goes through. There are also pores which can be opened or shut depending on how the proteins they are made of are oriented.

5.1.1.2 Facilitated Diffusion and Active Transport

Our bodies rely on a number of large, water-soluble molecules like amino acids and sugars. A glucose molecule is about 1 nm across – much larger than oxygen. So there are special means for large molecules to pass through membranes. *Facilitated diffusion* refers to a process that uses ion channels or carrier proteins to transfer specific molecules across the membrane. Ion channels are pore-forming proteins that sit in the membrane and allow a particular molecule to pass through to the other side. They can be open at both ends simultaneously. Carrier proteins can do a similar job, but transport through them is slower, and they only have the inside or outside end open at one time. A carrier protein can detect when the desired molecule has entered from the outside, and then close the outside end and open the inside to let the molecule pass into a cell, as shown in Figure 5.2.

As the name implies, for facilitated diffusion to occur, there must also be an appropriate concentration gradient across the membrane. There are also many

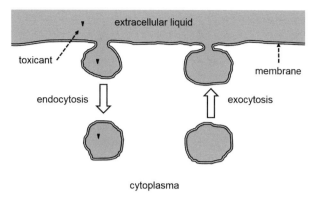

Endocytosis and exocytosis

biochemical situations where there is a need to transport molecules against the concentration gradient. For these situations *active transport* is used. Active transport is also based on carrier proteins (but not ion channels). For many molecules, the process operates against a concentration that is 30 to 50 times higher on the delivery side of the membrane, but in some cases it can be much higher. Unfortunately, it is also possible for a carrier protein to be tricked into helping a toxicant that has some chemical similarity to a desirable cell input. For example, the active transport mechanism for calcium can facilitate lead transport through membranes – the physiochemical characteristics of these ions are similar enough in this case.

5.1.1.3 Endocytosis and Exocytosis

Another way for something to pass into a cell is for the membrane barrier to completely envelop it. In *endocytosis*, the membrane develops an inward-pointing pocket or 'vesicle' that eventually detaches from the membrane and goes into the cell. These typically range in size from tens of nanometres to several micrometres in diameter – see Figure 5.3. The opposite process, for expelling material, is called *exocytosis* – also shown in Figure 5.3. These processes are useful in many circumstances; for example, neurotransmitter substances can be stored in vesicles ready for use, allowing rapid release of the neurotransmitters across synapses when a signal has to be transmitted, rather than waiting for nerve cells to synthesise them. Another key feature of these processes is that they can handle relatively large molecules that have to pass through membranes. A downside is that when the vesicles are large, there is a correspondingly large potential for material other than the intended cargo to be passed through the membrane. So naturally these processes are relevant for the absorption of toxicants.

5.1.1.4 Dermal Transfer

Your skin is a complex, multilayered membrane and is your outer defence against toxicants, infection and dehydration. The outer layer of skin, the epidermis, can be divided into five layers, but for simplicity only the two thickest ones are shown in

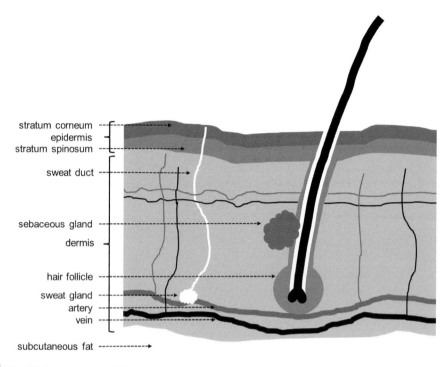

stratum corneum
epidermis
stratum spinosum

sweat duct

sebaceous gland

dermis

hair follicle

sweat gland
artery
vein

subcutaneous fat

Figure 5.4   Simplified cross-section of skin

Figure 5.4. These are the *stratum corneum* on the outside, and the lower *stratum spinosum*. Ninety-five per cent of the epidermis consists of cells called keratinocytes and they are constantly replacing themselves, migrating from the bottom of the stratum spinosum to the surface. As they reach the stratum corneum, they become increasingly tough as they fill with the protein keratin. The outer surface of the epidermis is constantly being worn off, so the epidermis replaces itself entirely about every 14 days.

The epidermis is typically between 0.4 and 1.5 mm thick (Chu et al., 2003). Its thickness determines the rate at which toxicants can diffuse into the body. The thickest places are the palms of your hands and the soles of your feet, while the epidermis on the back of your hands and your eyelids is relatively thin to allow easy movement of those parts of your body. Consequently, it is much easier for chemicals to enter your body through those regions. As indicated in the figure, the epidermis is punctured with sweat glands and hairs, but since these represent less than 1% of the outer surface area of the skin, they are unimportant as transfer sites compared to the rest of the surface. Once a toxicant has managed to get through the epidermis, it confronts a network of fine capillaries that supply the upper edge of the dermis with blood. So at this point a toxicant can easily enter your bloodstream. The network of capillaries also supplies moisture to the epidermis and the sweat glands. Additional moisture increases the permeability of the skin. This is something to think about in relation to the use of non-breathable gloves as a safety precaution in industries using chemicals. The gloves may provide an extra protective layer that reduces the rate at which an organic toxicant reaches the epidermis, although they are not entirely impermeable to lipophilic toxicants. But on the other hand, if they make you sweaty,

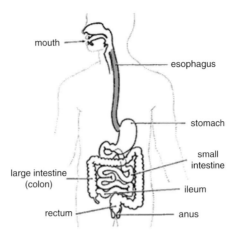

The human digestive tract

they may increase the rate at which any toxicant passes through the epidermis, paradoxically raising your absorption rate. Activities that involve the wearing of rubber or plastic gloves while handling lipophilic toxicants should therefore be limited to short periods of time.

5.1.1.5 Oral Transfer

The human digestive tract is like a 5-m-long hose through the body, but even though that may sound long, since it is not smooth, its surface area is much larger than this number would suggest. The small intestine alone has an interior surface area of about 30 m^2 (Helander & Fändriks, 2014) – which is the floor area of many studio apartments. This is an order of magnitude smaller than some older surface area estimates, but still very large and consistent with the digestive tract's primary purpose, the absorption of useful biochemicals and ions. As you would expect from the previous discussion, the walls of the digestive tract are loaded with active transport proteins. So considering its character and surface area, the digestive tract presents a large opportunity for toxicant fluxes if you ingest contaminated food or liquids.

   It takes about 24 hours for food to travel the twisting path from your mouth to your anus, as shown in Figure 5.5. Along the way, your body transforms food physically and chemically. This starts with the grinding processes in your mouth that decrease the particle size of the food, increasing its surface area and thus the potential for chemical reactions at the surface of the particles and mass transfer into the liquid phase. After travelling down the oesophagus (or, in North America, the esophagus), food reaches the stomach – a particularly acidic environment. Your body excretes hydrochloric acid into the stomach, reducing the pH of its contents to about 1. Downstream, the pancreas excretes an alkali solution that raises the pH back to about 5.3. These changes are important when it comes to the absorption of chemicals with acidic or basic functional groups, as shown in Figure 5.6. In this example, food is contaminated with benzoic acid and aniline (a weak base). In the stomach, these two molecules respond to the acidic environment in ways that influence their hydrophobicity. Benzoic acid has an acid dissociation constant (pK$_a$) of 4.2

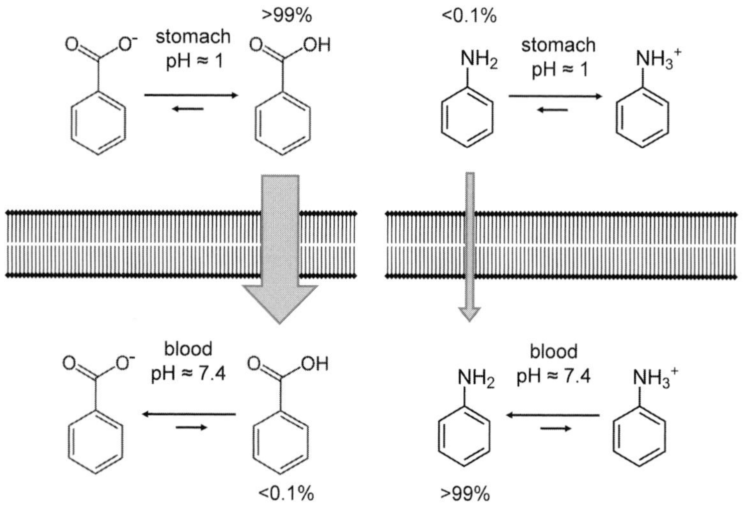

Figure 5.6   Influence of pH on absorption of benzoic acid and aniline

(Aylward & Findlay, 1974), which means that at a pH of 4.2, half of the acid is in each of the two forms shown in the figure, but at pH 1, the equilibrium favours the fully protonated form on the right. So more than 99% of the molecule will be in this relatively lipophilic form in the stomach. Meantime, any benzoic acid that diffuses to the other side of the membrane will transform back to the deprotonated, water-soluble form on account of the higher pH in the human blood (7.4). So the different pH conditions on either side of the membrane drive the equilibria in opposite directions, fostering the existence of a strong concentration differential across the membrane for benzoic acid.

The opposite is true of aniline. It has a pK_a of 4.6, which is almost the same as that of benzoic acid, but in this case, it refers to the equilibrium between aniline and its conjugate acid, as shown in the figure. So in the stomach, the majority of this compound is found in the protonated, hydrophilic form, which is less able to pass through the hydrophobic membrane. Furthermore, in the blood on the other side of the membrane, the equilibrium favours the lipophilic, uncharged form of the compound. So compared to the benzoic acid case, the total concentration of aniline (protonated and unprotonated) would have to be much higher in the stomach for a concentration difference to drive diffusion into the blood.

Since the pH rises to about 5.3 in the small intestine, these equilibria more than reverse in this part of the digestive tract, putting the vast majority of aniline into its unprotonated, relatively lipophilic form. This fact, coupled with the relatively long residence time of food in the small intestine compared to the stomach, means that we can absorb this toxicant easily in our digestive tracts.

5.1.1.6 Respiratory Transfer

The other parts of the human anatomy that are particularly designed for absorption are the lungs. As shown in Figure 5.7, air passes down the trachea and the flow is distributed by the bronchi to about 400 million alveoli. These are microscopic air sacs

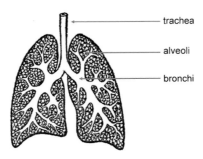

trachea

alveoli

bronchi

Figure 5.7   Human lungs

surrounded by tiny capillaries. The total area of these alveoli is estimated at about 50 to 75 m^2. So from this point of view, the lungs are your premier absorption organ, and they are the typical way toxicants are absorbed in the workplace.

Airborne toxicants can enter the lungs in all three phases: gas, solid and liquid. (Aerosols can consist of solid particles or liquid droplets.) The extent to which we absorb them depends on several factors. A key factor controlling whether you absorb an airborne contaminant is particle size. When exposed to dust, smoke or aerosols, the largest particles (above about 5 μm diameter) will generally be stopped in your nasal cavity, mouth or throat. The momentum of such particles and the relatively rapid airflow causes collisions between them and your mucous membranes, to which they adhere. Smaller particles (1–10 μm) are deposited in the bronchial passages, while particles below 1 μm will generally make it all the way to the alveoli.

Another key factor is water solubility: your airways are kept moist from your nose all the way into the alveoli, so if a toxicant is water soluble, it will pass into solution more easily. If the concentration of the toxicant in the air is low, it may be removed from the air by dissolving into the moisture inside the nasal cavity. There are plenty of nerve cells here, so if a highly soluble gas like sulphur dioxide (SO_2), hydrogen chloride (HCl) or ammonia (NH_3) is present in the air, a change in pH will be rapidly detected and lead to irritation and coughing, which may provide you enough warning to escape the situation. In principle, it is possible to absorb a toxicant molecule anywhere through the moist mucous membranes (or 'mucosa') that line your respiratory system from the nose and the alveoli, but the surface area of the respiratory system between the nose and bronchi is very small in comparison to the alveoli. At higher concentrations, of course, the nasal filtration mechanisms will not remove all the toxicant from the air you breathe and the contaminants will reach deep into the lungs where fewer nerve cells are present and the absorbing surfaces are large.

Gases like isocyanates (R-N=C=O, where R is an alkyl group), chlorine (Cl_2) and ozone (O_3) are less water soluble and can be expected to reach the bronchi. The mucus layer on the inside of the bronchi is thinner than it is higher up in the respiratory system, which makes absorption into the bloodstream easier. The bronchi will typically react to the inflammation this causes by constricting, which can cause breathing difficulties and chest pain. Water-insoluble gas-phase contaminants will reach the alveoli, where there are no nerve cells and the physical conditions for transfer into the bloodstream are excellent.

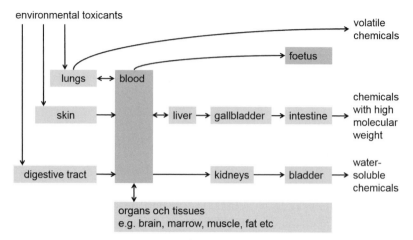

Figure 5.8   Overview of distribution pathways for toxicants

The rate at which you pump air into the lungs is about 5 L/minute when resting, but during physical work, this can quadruple to 20 L/minute. On the other side of the alveolar membranes, blood is pumped into the lungs at about 5 L/minute, or 10 L/minute during physical work. This is one of the reasons authorities recommend people stay home during industrial air pollution events. Depending on the geography, it may not be that the air quality in the home is much better than outside, but the potential for toxicants to enter the lungs and subsequently the bloodstream is minimised if people are resting.

5.1.2 Distribution

As shown in Figure 5.8, the main way in which absorbed toxicants are distributed through the body is via the blood. Water-soluble toxicants may bind onto blood protein, which tends to make the blood a reservoir for them, slowing down the exposure of critical organs to the contaminants and allowing the body's metabolic and excretion mechanisms more time to expel the toxicants. On the other hand, blood reaches almost every tissue in the body, so equilibrium processes will potentially distribute the (reversibly) bound toxicants.

One example of how a toxicant can be distributed around the body is methanol. Methanol can enter the body by any of the means shown in Figure 5.8, though it most regularly happens via unregulated (sometimes illegal) alcohol supplies, when people think they are drinking ethanol (for example in vodka). Methanol is highly volatile, so when it enters the blood via the digestive tract, it will mostly leave via the lungs or urine, without being metabolised. This is a good thing, as we see in Section 5.1.3.2.

Other than blood, another key reservoir for contaminants is the fatty tissue in the human body. Fat is stored in many locations in the human body: under the skin, around organs, in bone marrow and in the breasts. Fatty tissues have important roles in your body, with functions including thermal insulation and electrical insulation (the myelin sheaths around some nerve cells in the brain and elsewhere). It also has other energy management roles, including heat generation ('brown fat') and energy storage ('white fat'). The latter is of particular significance in the toxicokinetics of migratory birds (and other animals which have adapted to natural climate variations by adding large quantities of fat at certain times of the year). They may slowly absorb toxicants by feeding in polluted environments and feel no ill effects until the sustained weight loss of a migratory voyage releases the fat-stored toxicants, exposing a target organ to an extreme dose. The target organ of an overweight person receiving a dose of a systemic lipophilic contaminant may experience a lower short-term exposure compared with the target organ of a thin person receiving the same dose, on account of the contaminant's distribution to the fatty tissues. On the other hand, if the contaminant is stable, the target organ of the overweight person will be exposed to the contaminant for a longer time.

Bone tissue can similarly act as a reservoir for contaminants like strontium and lead, as they are able to replace calcium. Lead has no toxic effect on the bone, but can be remobilised from the bones (where it has a half-life of about two years) and affect the functioning of the blood, nervous system, kidneys and other tissues.

Two important barriers to the distribution of toxicants are worth mentioning here: the blood-brain barrier and the placental barrier. The endothelial cells, the ones at the very surface of a capillary, are jammed together in an unusually tight arrangement in the brain. This barrier allows the passage of water and some gases and lipid-soluble molecules by passive diffusion, as well as the selective transport of some molecules that are crucial to neural function such as glucose and amino acids. The blood-brain barrier is thought to be an important protection against bacteria and toxins that humankind was routinely exposed to during our evolutionary history. However, like with most barriers, there is some potential for it to be crossed by certain toxicants. One interesting example is mercury. As a metallic ion (Hg^{2+}), it is water soluble and it is not likely to affect the brain directly. On the other hand, methylmercury salts, like methylmercury chloride (H_3CHgCl), are fat soluble and can pass through the blood-brain barrier. Worse: on the other side, they can oxidise to the metallic ion form, thus reducing the likelihood of elimination from the brain, where it can degrade myelin.

The placental barrier has historically been thought of as an important protection for the unborn, and indeed it plays an important role in separating the placental and parental bloodstreams. However, this barrier is permeable compared with the blood-brain barrier, as it has to allow the transfer of oxygen and nutrients to the foetus and the removal of carbon dioxide and waste products (e.g. urea) back to the mother. This is accomplished with a wide variety of active and passive membrane transport proteins. There is also some metabolic activity in the placenta (Coleman, 1986), which may change the toxicity of contaminants, as described in the next section.

5.1.3 Metabolism

Metabolism (or 'biotransformation') refers generally to the range of chemical transformations performed within the cells of living beings. Metabolism converts food into the body's own energy carriers, or into building blocks like proteins, nucleic acids and lipids. It is also important for the elimination of wastes. When it comes to toxicology, metabolism is an important step in the excretion of toxicants, and can also decrease (or, in some cases, increase) the hazard posed

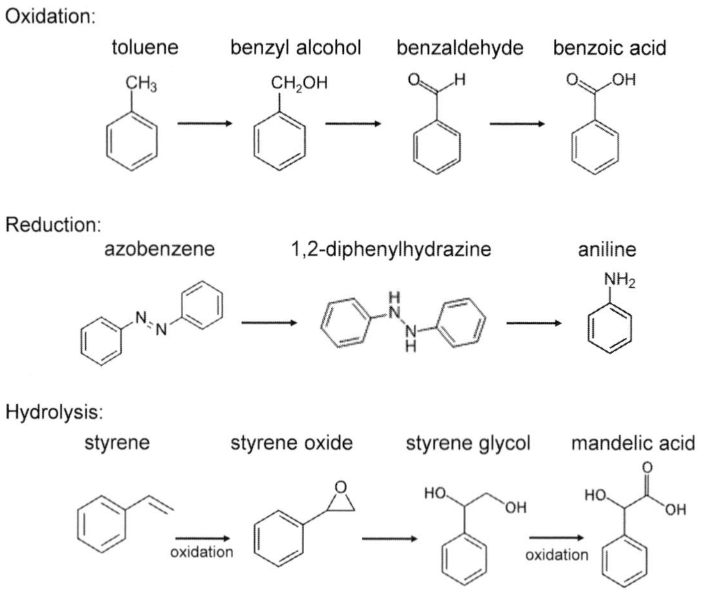

Examples of phase 1 reactions

by different molecules. The overall process is about trying to make the toxicant more water soluble. The metabolism of toxicants is usually considered as a two-phase (or two-step) process.

5.1.3.1 Phase 1: Adding a Useful Functional Group

There are many possible kinds of phase 1 reactions. Three key classes are oxidation, reduction and hydrolysis. In the case of oxidation reactions, these may be catalysed by enzyme systems. Some examples of phase 1 reactions are shown in Figure 5.9. Note that one could also call the first step of the oxidation of toluene a hydroxylation reaction (Hanrahan, 2012). Note also that in these cases, several of the intermediate products of the metabolic process are readily water soluble. For example, the mandelic acid shown in the figure can be readily excreted in urine, so conjugation in phase 2 is unnecessary.

5.1.3.2 Phase 2: Conjugation with a Water-Soluble Group

There are a number of possible phase 2 reactions, all of which add an extra water-soluble structure to the product of the phase 1 metabolism. Like mandelic acid in the previous example, benzoic acid (the product of the phase 1 oxidation of toluene) is also reasonably water soluble (3.4 g/L at 25°C), but when converted to hippuric acid (Figure 5.10) by conjugation with the amino acid glycine, it is a little more soluble (3.8 g/L at 25°C).

As a consequence of this metabolic pathway for toluene, the presence of hippuric acid in the urine has been used as a diagnostic indicator of human exposure to toluene (Hanrahan, 2012).

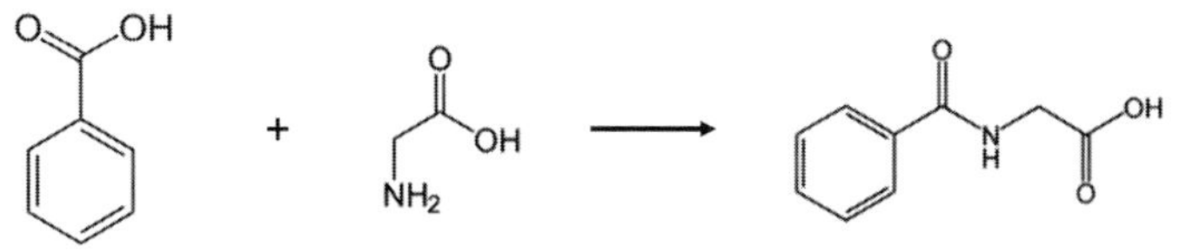

Figure 5.10 Conjugation of benzoic acid with glycine

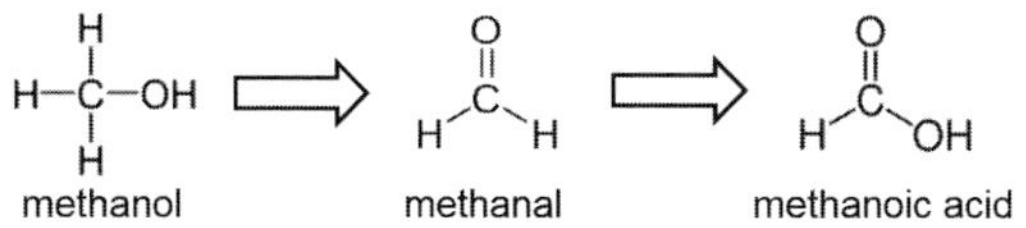

Figure 5.11 Oxidation of methanol

Other key examples of phase 2 reactions include conjugation with glutathione, glucuronic acid, sulphate and amino acids. The first two of these are perhaps the most important. Glutathione typically attaches via the thiol group, while the glucuronic acid reaction has to be catalysed by a prior enzymatic reaction with uridine diphosphate.

The metabolism of methanol is a simple example of when metabolism can cause more problems than it solves. As shown in Figure 5.11, methanol is enzymatically oxidised by alcohol dehydrogenase to methanal ('formaldehyde') and then to methanoic ('formic') acid. The lethal dose of methanal is relatively low (0.10 g/kg is the oral rat LD_{50} – see Chapter 6 for an explanation of this term) compared with methanol (5.6 g/kg) and methanoic acid (1.1 g/kg). Methanal attacks the optic nerve in humans, thanks to the naturally high concentration of alcohol dehydrogenase in that part of the body. The good news for victims of methanol poisoning is that this enzyme has a higher affinity for ethanol. So part of the treatment for methanol poisoning is to administer small doses of ethanol to the patient. This ethanol preoccupies the enzyme, providing more time for the methanol to be removed via the lungs and urine.

5.1.4 Excretion

On any normal day, the main ways you can excrete toxicants are via your respiratory system, your urine or your excrement. Women of childbearing age have two special, additional means: by giving birth and expressing milk. As mentioned previously, the placenta is relatively pervious to toxicants that have found their way into the mother's blood, so the elimination of the placenta and the delivery of a child are relevant to maternal excretion of contaminants. Indeed, it is likely that some persistent organic toxicants have been passed down through several generations. Human breast milk contains nutritionally valuable quantities of fat (about 4% – more than typical commercial cows' milk), so expressing it is an effective way for the mother to get rid of lipophilic contaminants like polychlorinated biphenyl (PCB) compounds and methylmercury. Calcium is also an important part of breast milk (30 mg/L – about a third of the concentration in cows' milk), so lead can also be expressed with it. Of course, if the milk is expressed during breastfeeding, this transfer of contaminants is unwanted, but the weight of evidence suggests that for mothers outside highly contaminated environments, breastfeeding still provides their children with major immunological health benefits, despite PCB contamination (Mead, 2008).

5.1.4.1 Respiratory System

Mucus is a vehicle for the elimination of particulate contaminants and dissolved toxicants from the respiratory system. Your body continuously produces surprising quantities of mucus to line the surface of your airways – about a litre of mucus is produced in the nose each day. You do not see most of it because it is slowly moved to your throat by gravity and the tiny hairs (called 'cilia') that line your airways, and then unconsciously swallowed. Water-soluble contaminants and dust particles trapped in the mucus are thus rapidly carried out of the airways and into the digestive tract – their typical half-life in the major airways is a few hours. Alternatively, if you are exposed to cold air, the cilia may slow down or temporarily cease their rhythmic beating, causing a runny nose and carrying contaminants out that way. They can also leave via the mouth or nose, if you inhale irritating particles of sufficient quantity or size, and you instinctively react by coughing, sneezing or spitting out mucus contaminated with these particles.

Although most cells have some kind of cilia, this cleaning system does not operate in the alveoli. In these tiny spaces, a kind of endocytosis called phagocytosis is important. There are *macrophages*, a kind of white blood cell, which roam the surface of the alveoli and absorb particulate contaminants (Reynolds, 1985). The vesicles (or, in this case, 'phagosomes') formed via endocytosis by the macrophages can destroy the contaminant using enzymes and reactive peroxides, or the macrophages can carry the contaminant with them towards the smaller bronchi or the lymphatic system, where the cilia take over the transportation role. This overall process is slower than the removal process in the upper respiratory system, with a half-life of about two months (Birgersson et al., 1995).

These systems are of course not absolute protection from all possible toxicants. As mentioned earlier, the alveoli are designed for absorption of gases, and many poisonings have happened in workplaces through respiratory exposure to toxic gases. Moreover, some kinds of particles are almost impossible to expel using the body's natural defences. Some kinds of microscopic particulate matter, like silicon dioxide or the more complex silicate mineral asbestos (Figure 5.12), are extremely sharp-edged and can destroy the macrophages that try to remove them. A slow process of lung irritation, accumulation of fibrous tissue around the particles and possibly physical interference with the DNA in cells can lead, over decades, to reduced breathing capacity, cancer and an early death. Silicosis and asbestosis are incurable. Engineered and legislative protections (in other words, physical and legal barriers to exposure) are the key defences from these hazardous particles.

5.1.4.2 Excrement

The liver has a central role in your body's management of toxicants. In addition to its role in protein synthesis and the production of a wide range of biochemicals, it is the place where important metabolic processes for the elimination of hydrophobic

Figure 5.12 Scanning electron micrograph of asbestos fibres. Source: Wikipedia

contaminants operate – both phase 1 and phase 2 as described earlier. It is directly connected to your digestive tract via the *portal vein*, which gives us the evolutionarily valuable capacity to remediate contaminants reaching the blood from the digestive tract, before the blood gets sent to the heart and other parts of the body.

The cells in your liver contain relatively large pores to allow the absorption of proteins into the cells, so filtration processes are important here and contaminants absorbed onto blood protein are taken in. Liver cells produce an alkali surfactant called *bile* (or 'gall') which helps to emulsify fats in the digestive tract, to allow digestive bacteria to do their work. Bile also has the important function of transporting metals like mercury, lead and arsenic, as well as the heavier organic toxicants (molecular weight above 500 grams per mole) out of the liver. Bile is drained from the liver in various bile ducts, and stored in the gallbladder for release into the digestive tract, just below the stomach where the small intestine begins. Bile salts like cholic acid are then actively or passively absorbed into the wall of the small intestine and thus returned to the liver.

You may think the plumbing in this system sounds at risk of contradicting itself, and you would be right. But recycling bile rather than making it from scratch makes sense for a person untroubled by toxicants – bile salts may recirculate 20 times before excretion (Kuntz, 2008). For the most part, the toxicants released into the intestine via the bile duct have been metabolised into their water-soluble, phase 2 products, and therefore absorption downstream in the small and large intestines is minimal. They therefore leave the body via the anus. On the other hand, some of the conjugates created by metabolism in the liver can be split by microorganisms in the small intestine, returning toxicants to their lipophilic forms, which are then readily absorbed back into the bloodstream and returned to the liver via the portal vein. This *enterohepatic circulation* is a problem for example in relation to methylmercury.

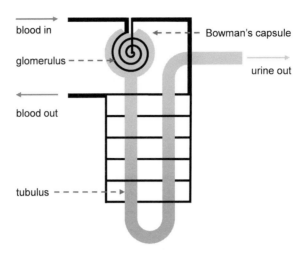

Figure 5.13   Simplified representation of a nephron

When grain treated with a methylmercury fungicide poisoned thousands of people in Iraq in 1971, the victims were administered a polythiol resin (a polymer containing many SH groups) via the mouth in an attempt to make the resin (particularly the sulphur in it) bond with the methylmercury in the digestive tract and to prevent enterohepatic circulation and redistribution to other organs.

5.1.4.3 Urine

Your kidneys have the important role of regulating the pH of your blood, your blood pressure and your electrolyte balance. Each of your kidneys is equipped with a million *nephrons*, the tiny filtration units that produce urine. Blood flowing into your kidneys is distributed to the nephrons and their tiny networks of capillaries, known as *glomeruli* (each nephron has one 'glomerulus' – see Figure 5.13). A glomerulus is surrounded by a filtration unit called a *Bowman's capsule* (named after the man who discovered it in 1842), which has relatively large pores. Blood passes into your Bowman's capsules at a high rate – about 125 mL/minute – which means your blood *plasma* is filtered about 60 times per day. The filtrate leaving the capsule is very similar to plasma, but large proteins are retained in the glomerulus, as are blood cells. On the other hand, water, glucose, salt, urea and amino acids pass freely into the capsule. So any toxicants absorbed onto proteins are retained, but water-soluble metabolites of toxicants and valuable water-soluble substances pass into the urine.

To get the valuable substances back into the blood, the urine travels through a convoluted piece of plumbing called the *tubulus*, which is nestled in another network of capillaries. The tubulus is equipped with various active transport systems to ensure that glucose, amino acids and some other valuable substances are reabsorbed. The regulation of pH involves the excretion of hydrogen ions and a reversible mechanism by which an enzyme called *carbonic anhydrase* strips hydrogen and oxygen off bicarbonate on one side of the membrane, to make (smaller) CO_2 molecules. These are able to diffuse through the membrane, after which the same kind of enzyme can reconstruct the bicarbonate. Water is also mostly reabsorbed, leaving up to 2 L/day (1.4 mL/minute) of urine to continue on to the bladder and elimination from the body. Since the urine is more concentrated, any lipophilic substances present in it can more easily diffuse back into the body, but water-soluble contaminants are unlikely to return unless they mimic the appearance of some substance with an active transport mechanism for reabsorption in the tubulus. There are also some active transport proteins working to move a few substances out of the blood and into the urine in the tubulus.

5.2 Toxicodynamics

A toxicant can cause you harm in many ways. We like to distinguish between local and systemic effects, and also between *acute* and *chronic* effects. The latter terms distinguish between effects that occur almost as soon as you are exposed to the toxicant (e.g. intoxication by alcohol) and those that may take much longer, perhaps years, to cause perceptible harm (e.g. silicosis). In this book, a few key classes of toxicant effects are described: impacts on energy conversion and on the nervous system, allergenic effects and genetic impacts.

5.2.1 Energy Conversion

The citric acid (or 'Krebs') cycle is a key metabolic pathway in aerobic lifeforms like us. Any good textbook on biochemistry should provide a detailed description of this complicated sequence of reactions driven by eight different enzymes (e.g. Chatwal & Powar, 2007; Hardin & Knopp, 2013). For the purposes of this book, we need a brief overview of some of its inputs and outputs to help us to understand how toxicants interfere with it.

If we simply burn the monosaccharide glucose in oxygen, we generate an impressive amount of heat:

$$C_6H_{12}O_6 + O_2 \rightarrow 6\ CO_2 + 6\ H_2O + \text{heat}\ (2{,}805\ \text{kJ/mol})$$

This may be useful, but it is overkill for many of the chemical reactions that our bodies need to operate. Instead, one of the main reactions our bodies use to deliver energy is the liberation of a phosphate ion from adenosine triphosphate (ATP) (see Figure 5.14).

By shortening the triple phosphate chain in ATP in this way, adenosine diphosphate (ADP) is generated and a relatively modest 30.5 kJ/mol is produced. This reaction is

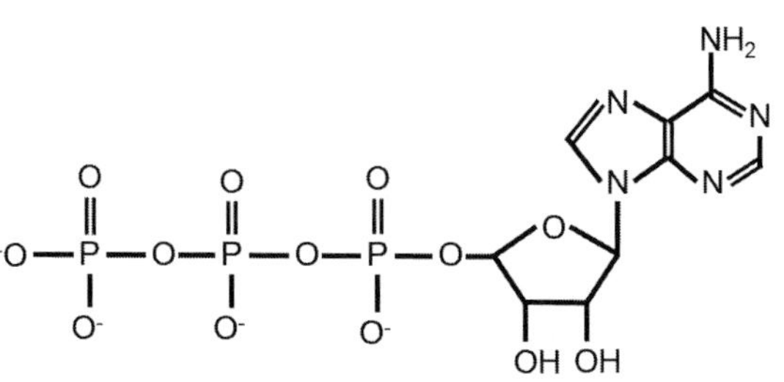

Figure 5.14   Structure of ADP

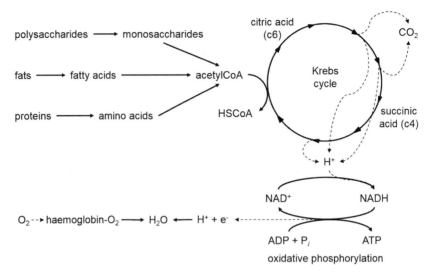

Figure 5.15   Overview of metabolic energy production

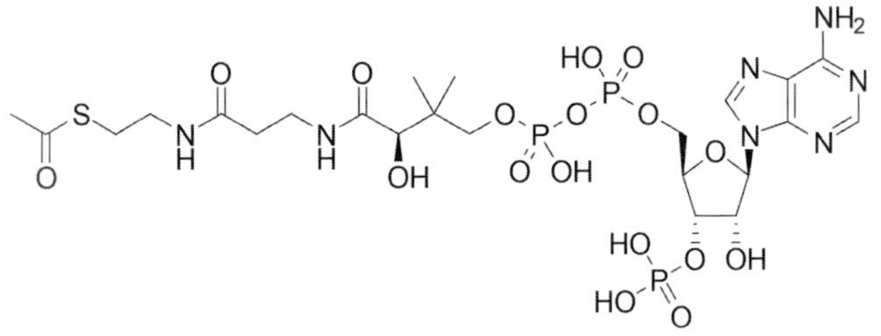

Figure 5.16   Acetyl coenzyme A

extremely useful inside our living cells, and it has the practical characteristic of being readily reversible. This leads to ATP being called 'the molecular unit of currency' of intracellular energy transfer (Knowles, 1980). So how do we produce our ATP?

As shown in Figure 5.15, the first steps in our energy supply are the conversions of polysaccharides, fats or proteins into smaller building blocks. This activity occurs in the liver. These monosaccharides, fatty acids and amino acids are then subdivided further until they are units of only two carbon atoms, and then bonded to *coenzyme A* to form *acetyl coenzyme A* (Figure 5.16; called acetyl-CoA in Figure 5.15). Delivery of the acetyl-group shown at the left of that figure into the Krebs cycle creates citric acid, which contains six carbon atoms. This is converted stepwise into (four-carbon) succinic acid, with the liberation of carbon dioxide and protons. In turn, the ionised form of an enzyme called *nicotinamide adenine dinucleotide* (NAD^+) is protonated by removing these protons from the Krebs cycle. The protons in NADH are then used in another multi-stage enzymatic process called oxidative phosphorylation, which produces ATP. During oxidative phosphorylation, electrons are donated to oxygen, giving it a negative charge (O_2^-). This oxygen ultimately reacts with protons derived from the Krebs cycle, making water.

By breaking down glucose carefully, our metabolism can eventually generate 32 molecules of ATP from a single glucose molecule. Of course, this level of complexity leaves the metabolic process open to toxicological attacks on many fronts. Some toxicants can attack the metabolic process near the top of Figure 5.15. For example, chloroacetic acid and fluoroacetic acid easily penetrate mucous membranes and skin, and bind with coenzyme A. This creates a product that resembles acetyl coenzyme A to such an extent that instead of producing citric acid, the responsible enzyme makes a halogenated version of it. This cannot be handled in the next steps of the Krebs cycle, and even if other citric acid is present, the halogenated citric acid prevents the other enzymatic conversions from taking place. The central nervous system and the heart are the first impacted by this brake placed on the Krebs cycle (Birgersson et al., 1995).

Cyanide and hydrogen sulphide impact energy production further down the metabolic chain, by bonding to one of the iron-containing enzymes that is part of the oxidative phosphorylation process shown in Figure 5.15. In this case, even when the oxygen supply from the lungs to the cells is adequate, it cannot be used. This can lead to unconsciousness, breathing difficulties and death. Hydrogen cyanide gas at a concentration of 100–200 ppm will kill a human in 10–60 minutes – it was used for genocidal purposes by the Nazis during the Second World War and is still employed for legal executions in three states of the United States. One treatment approach for cyanide poisoning is (paradoxically) to oxidise the iron in the blood (to ferric (3^+) iron), which then helps the cyanide to bond with the haemoglobin, preventing it from entering the mitochondria in the cells and messing with oxidative phosphorylation.

One can also 'stop the motor by blocking the exhaust pipe'. The flow of oxygen to cells is needed to remove excess electrons generated in metabolism, thus the oxidative phosphorylation process can be retarded if the supply of oxygen from the lungs is inadequate, causing the production of ATP to be interrupted. Many toxicants can cause this interruption. For example nitrite (NO_2^-) can react with the ferrous (2^+) iron in haemoglobin, oxidising it to ferric (3^+) iron, which prevents it from bonding to oxygen gas in the lungs. Early symptoms of this form of poisoning include blue lips and breathlessness. Some parts of the body are more resistant to oxygen deficiency, for example muscles can operate anaerobically for a while, but the nervous system is more sensitive and can be rapidly damaged when sodium and calcium transport proteins lose power, resulting in an accumulation of these ions and water. Cerebral oedema (accumulation of liquid in the brain) due to a lack of oxygen can cause permanent brain damage. Carbon monoxide gas (CO) poisoning operates in a different way: the CO bonds with the haemoglobin at the sites that should bond with the oxygen, preventing the oxygen from bonding to the haemoglobin. Oxygen gas is used in the treatment of CO poisoning to try to accelerate the removal of CO and to increase oxygen uptake.

5.2.2 Nervous System

Nerve cells ('neurons') connect various parts of your body to the brain. They are typically about a micron (μm) in diameter, but range in length from about a millimetre to more than a metre (in the sciatic nerve, from the base of your spine to your toes). Neurons are surrounded and insulated by a fatty 'myelin' sheath. Signals can be passed from the end of one neuron (the presynaptic neuron) across the tiny gap

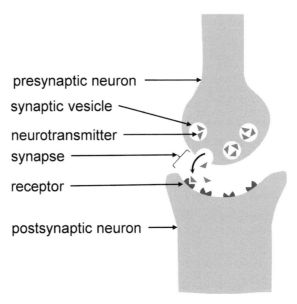

acetylcholine   choline   glutamate

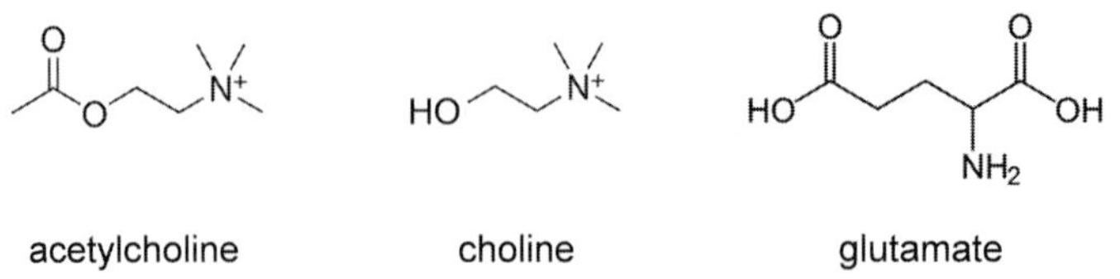

Figure 5.17   Common neurotransmitters

presynaptic neuron

synaptic vesicle

neurotransmitter

synapse

receptor

postsynaptic neuron

Figure 5.18   Simplified diagram of a synapse

('synapse') to another (the postsynaptic neuron) by either electrical or chemical impulses. The latter is the more common method, and a wide diversity of neuro-transmitting chemicals exists, which of course leaves the nervous system open to interference by a wide range of toxicants.

Two of the main neurotransmitters are glutamate and acetylcholine (Figure 5.17). Serotonin and dopamine are other well-known neurotransmitters. Glutamate is the most important neurotransmitter in the human nervous system, accounting for more than 90% of synaptic connections in the brain. Acetylcholine is (among other synaptic roles) the neurotransmitter that is used at the junction between nerves and muscles.

Neurotransmitters have to pass rapidly from one nerve cell to the other – we would be rather slower if the process of neurotransmitter synthesis had to be carried out every time an impulse reached the end of the nerve. Instead, the neurotransmitters are sitting in synaptic vesicles ready to be released via exocytosis into the synapse – see Figure 5.18. In the case of acetylcholine, an enzyme called acetylcholinesterase is present on the surface of the postsynaptic neuron, and it deactivates the acetylcholine by converting it to choline (Figure 5.17 again). This is important because it enables the receptor to be prepared for the next time a signal crosses the synapse. The choline

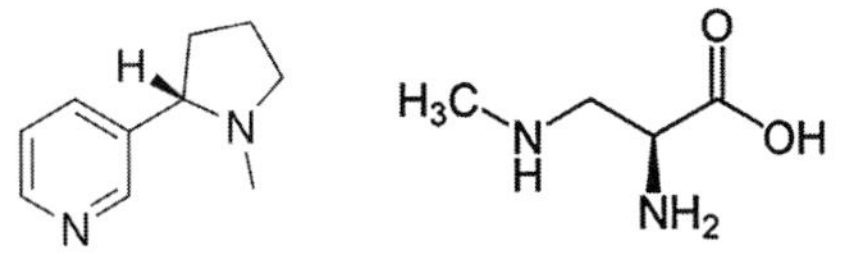

Figure 5.19   Nicotine (left) and *beta*-methylamino-L-alanine (right)

is subsequently reabsorbed by the presynaptic neuron. The process is similar for glutamate, but the deactivating enzyme is called glutamate dehydrogenase.

Toxicants can affect the nervous system in many ways. One problem concerns the myelin sheath. Lead has a wide range of toxic effects on the nervous system (Lidsky & Schneider, 2003), and one of them is the degradation of the myelin sheath, leading to short-circuiting between adjacent neurons and slower transmission along the neuron. Toxicants can cause too little or too much of a neurotransmitter to be released by the presynaptic neuron. An example of the latter is DDT. As its primary pesticidal action, it makes the presynaptic neuron more permeable to sodium ions, resulting in excess signalling. Toxicants can also mimic neurotransmitters and confound the post-synaptic neuron, and they can prevent removal of the neurotransmitter after it has passed its message across the synapse.

Nicotine in cigarette smoke is an example of a toxicant that can impact processes at the synapse by mimicking a neurotransmitter. This chemical is sufficiently similar to acetylcholine that binds to the same receptors, blocking the acetylcholine signals but stimulating the postsynaptic neuron. 'Agonist' is a label for a neurotransmitter-mimicking chemical like this. In turn, other neurotransmitters are released, including dopamine, which is a part of your body's reward signalling system. (This is one of the factors that causes the addictiveness of tobacco smoking.) At low doses, nicotine can cause feelings of alertness, but it has a sedative effect at higher doses.

Beta-methylamino-L-alanine (BMAA) is another neurotoxin. BMAA is produced in cyanobacteria (blue-green algae), a common contaminant in untreated water supplies. BMAA is not fully understood, but it appears to be an 'excitotoxin' of the glutamate receptors. This means that it excessively stimulates the receptors like the presence of excess glutamate. High concentrations of calcium ions can then enter the neurons, activating a range of enzymes that go on to damage membranes and DNA.

Organophosphate is an example of a toxicant that prevents the removal of used neurotransmitters. Organophosphates denature the acetylcholinesterase enzyme, pre-venting the removal of acetylcholine from the synapse, resulting in an excess con-centration of this transmitter in the synapse, and overstimulation. When it affects the junction between neurons and muscles, this causes weakness, fatigue, cramps and paralysis. When it affects the central nervous system, anxiety, headache, convulsions, depression of respiration and circulation, and coma can occur. Organophosphate poisoning is a common problem on account of the widespread use of organophos-phate pesticides, and it causes several hundred thousand fatalities annually.

5.2.3 Allergenic Effects

Normally, your immune system reacts to 'antigens' – the term for molecules that activate the system when your body is under a genuine threat. Allergies are essentially overreactions by the immune system to some kind of environmental stimulus. In this case, the allergic reaction may be just as problematic as the unwanted molecule, if not for society at large, then for a particular individual. A wide range of compounds is known to generate allergic reactions. These compounds ('allergens') include natural substances like animal fur, dust mite excrement and peanuts, but also a range of manufactured products, for example perfumes and products containing nickel, chromium (IV) and formaldehyde. These potential toxicants are obviously widespread, as are occupational health problems connected with allergies. Compared to other effects of toxicants, allergic reactions can arise when people are exposed to very low doses of an allergen.

Your immune system contains two basic types of cells: B and T cells. The B cells create special proteins called 'antibodies' that float about freely in the blood. Antibodies identify and disable bacteria and viruses. There are millions of kinds of antibodies – if a B cell recognises a particular threat, it changes into a plasma cell and pumps out many copies of the right antibody to attack it. T cells operate differently, and are able to manage viruses hidden inside infected cells. Infection changes the outer cell membrane of an infected cell. A receptor in a T cell membrane recognises the change, and subdivides into many copies of itself. Each of the copies can attack and destroy cells with that infection.

To stimulate the immune system, a substance must be relatively large – having a molecular weight of at least 5,000 g/mol. Most chemical allergens are smaller than this, but they have the capacity to modify molecules naturally present in your body, so that these molecules in turn appear foreign to your B or T cells. For example, nickel weighs 58.7 g/mol and is probably a human carcinogen, but its allergenic effects are a problem for more people than the cancers it is suspected of causing among nickel smelter workers. Nickel is picked up by 'dendritic cells', a kind of cell designed to present antigens to the T cells. An excess of T cells is produced, causing a rash or eczema to develop in the skin. This 'contact allergy' may take a couple of days to develop, delaying detection. It may then take several weeks to subside. Once thus sensitised, there is no cure, so many people have become sensitive to nickel (more than 10% of the population, typically more women than men, presumably due to exposure to jewellery). On the other hand, persons with a nickel allergy can attempt to reduce their exposure, which is a reason the Swedish government has been attempting to eliminate nickel from Swedish and European coins (Sveriges Riksbank, 2011).

5.2.4 Genetic Effects

You will hopefully recall from high school that your genetic material is based on the sequence of pairs of nucleotides (adenine hydrogen-bonded with thymine; guanine hydrogen-bonded with cytosine) linked by two lines of phosphate deoxyribose polymer – generating the double-helical strands of 'deoxyribonucleic acid', or DNA. Mutagenic substances damage DNA in some way. Many substances are mutagenic, including oxidising or alkylating chemicals. High-energy electromagnetic radiation, for example X-rays and UV light, are also mutagenic. Mutagens can react with a nucleotide and convert it into another. They can also attack the DNA's polymer backbone and break it open. Benzo(a)pyrene is the main carcinogen in tobacco smoke. As shown in Figure 5.20, it is firstly activated by enzymatic processes

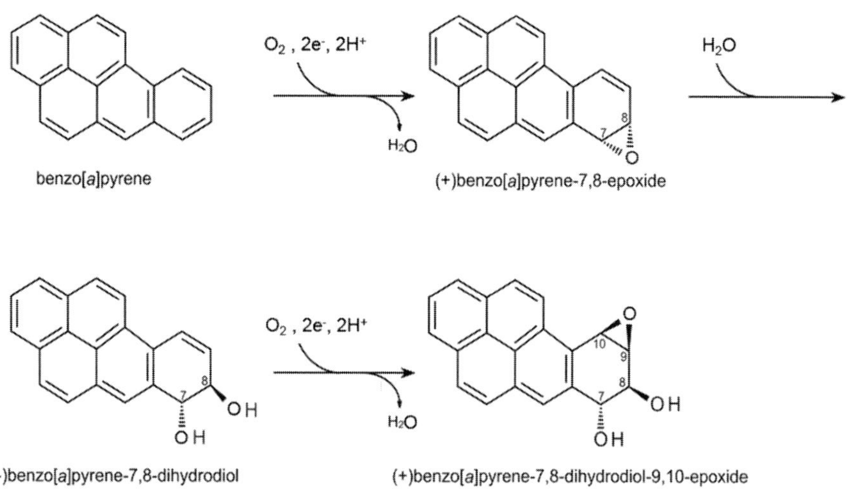

Figure 5.20   Metabolism of benzo(a)pyrene

in the body, becoming a more reactive compound (benzo[a]pyren-7,8-dihydrodiol-9,10-epoxide). Then it bonds to the nucleophilic guanine molecules, inserting itself between nucleotide pairs. This distorts the DNA double-helix, causing transcription errors when the DNA is copied to make new cells.

Non-smokers are not spared the risk of carcinogens forming inside the body. The low pH in the stomach can degrade many compounds, and the products can react with other chemicals present. One example of this is the formation of nitrosamines, which can be produced by the reaction of secondary amines with nitrite in acidic environments. For example, N-nitrosodimethylamine (also known as dimethylnitrosamine) is a product of the reaction of dimethylamine with nitrite (which is called an 'N-nitrosation' reaction), as shown in Figure 5.21.

This reaction can occur in the human stomach given the presence of secondary amines in various foods, including fish, vegetables and juice, and the use of nitrite salts as preservatives in meat. Nitrite can contaminate the drinking water supply on account of the use of mineral nitrogen fertilisers in catchments. Dimethylnitrosamine is also a potential by-product of chlorination of recycled water (Andrzejewski et al., 2005). This chemical is a particular concern because of the evidence that it causes cancer in mammals. Furthermore, dimethylnitrosamine has also been implicated in Alzheimer's disease and diabetes (Tong et al., 2009).

Mutation is a natural process and one which your body is equipped to repair. Problems can occur when the rate or kinds of mutations exceed your natural DNA repair mechanisms. But such permanent mutations are not always a problem – five different general outcomes are possible. You could be lucky and have a *positive* mutation which improves the performance of a cell. This is how evolution proceeds,

Figure 5.21   Production of dimethylnitrosamine by acid catalysis

but positive mutations are rare enough that the evolution of a new species is a rarely observed event in a human lifespan (but it does happen – see Wirgin [2011] for an example of the evolution via mutation of a kind of cod resistant to PCB toxicants). Another more likely outcome is a *neutral* mutation, one which changes a base pair without affecting its activity, or which results in a gene that has a similar effect to the original. *Silent* mutations occur in unused parts of your genetic code. Each cell contains the genetic blueprint for the whole organism, so, for example, if the part of the DNA in a nerve cell that codes for the production of haemoglobin is damaged, it does not matter – nerve cells do not produce haemoglobin! *Negative* mutations will reduce the capacity of the organism to survive, and *lethal* mutations are enough to destroy the cell.

Cancer occurs when cells lose control of their reproductive rate, and subdivide at an excessive rate, producing tumours. Cancer is a consequence of a sequence of mutations. Traditionally, the overall process has been divided up into the processes of *initiation* and *promotion*. The former process is some kind of irreversible process involving a mutation, and the latter process accelerates the development of the cancer. For example, it has been observed that rats exposed to tobacco smoke after the development of cancer cells grow larger tumours faster than rats not exposed to the smoke. The irritation of the lungs contributed to the promotion of the cancer.

Cancer is a common health concern, but not the only kind of genetic damage that chemicals can cause. Damage to the DNA in germ cells (eggs and sperm) or the cells that produce them is particularly problematic because of the potential for all the cells in the offspring of the damaged germ cell to contain copies of the damaged DNA – the relative likelihood of a silent mutation is reduced. Foetal cells are a similar concern since they will differentiate into many different parts of an animal prior to its birth. The earlier the mutation occurs, the more likely a miscarriage will occur – the first trimester is of the most concern for developmental toxicity. Tobacco smoke, PCB compounds, mercury and nitrate ions are examples of potential toxicants causing birth defects.

There is also the potential for the expression of genes to be affected without actual damage to the DNA sequence itself. When these impacts are heritable, they can be called 'epigenetic' impacts of chemical pollution, but impacts on gene expression can also be limited to exposed individuals. Endocrine disruptors are of particular concern today. These chemicals may mimic natural hormones in your body, which are required only in minute concentrations for their function of switching the operation of certain DNA sequences on or off. Like nerve toxins, some endocrine disruptors can act as agonists (binding to hormone receptors and having the same effect as the hormone) while others act as antagonists (blocking the receptor and preventing the real hormonal signal from arriving). Examples of common endocrine disruptors include PCB compounds, DDT, phthalates (plasticisers), lead and cadmium.

5.3 The Bigger Picture

This chapter is an introduction to the ways in which contaminants enter the body, are transformed in it, impact it and are excreted. Readers interested in developing a deeper understanding of toxicology will find useful additional information in the standard text: *Casarett and Doull's Toxicology* (Klaassen, 2013). It is important to possess some knowledge of these matters in order to protect yourself and your colleagues in workplaces containing chemical hazards. The science of chemical risk assessment is built on a knowledge of how chemicals are transported in the environment, and their toxicological properties in humans and other living things. This is the subject of the next chapter.

5.4 Review Questions

1. What is the difference between toxicodynamics and toxicokinetics? What are the four key processes in toxicodynamics?
2. Define facilitated diffusion and contrast it with active transport.
3. How can the use of protective gloves worsen the transfer of lipophilic contaminants through the skin?
4. How does the pH change as food travels through our digestive system? How can that help or hinder transfer of contaminants into the blood?
5. Describe the metabolism of toluene using chemical structure diagrams.
6. A worker in your factory complains of breathlessness, fatigue and headaches. You suspect carbon monoxide poisoning. How can the gas cause these effects?
7. By what mechanism does nicotine act on the nervous system?

6 Qualitative and Quantitative Risk Assessment

6.1 Introduction

Risk assessment is a very broad idea. People make informal, unconscious risk assessments every day – deciding in any given moment how fast to drive your car involves considering the risk of an accident or a speeding ticket. But risk assessment is also a specific professional skill and a field of academic research. This means that a student of risk assessment needs to use certain terms rigorously. 'Risk is the chance, within a time frame, of an adverse event with specific consequences' (Burgman, 2005). Many different kinds of risks can be examined in a formal, written risk assessment. Just a few of them are: accident risk (assessed by actuaries for insurance companies); the risk that a criminal will re-offend; financial risk (for example, when a bank officer is deciding whether to give you a loan); project risks (for example, construction managers have to consider whether they can assemble all the materials and human resources needed at the right time). The discussion of different kinds of risk assessment could fill many books – since this book is about environmental management, in this chapter, we focus on the assessment of risks to human health and the environment. Risks associated with chemicals and microbes are in focus. Typical situations in which the engineering professional may be required to perform or procure a risk assessment include the assessment of treated wastewater discharges into a river, target setting for contaminated land remediation and the evaluation of a new consumer product or a new industrial facility.

Depending on where you work and the kind of risks you have to manage, you will find different guidance documents that are relevant to the challenges you face. Some of the relevant international standards are mentioned here. ISO31000 'Principles and Guidelines on Implementation' is a document that addresses the entire management system for the design, implementation, maintenance and improvement of risk management processes. In general terms, this provides an institutional perspective on risk assessment. More technical detail about the actual process of assessing a risk is offered in ISO31010 'Risk Management: Risk Assessment Techniques'. Depending on the kinds of risks you face in whatever industry you work in, different guidance documents will be relevant. For example, regarding chemical risks, the European Union has its own very detailed 'Guidance on Information Requirements and Chemical Safety Assessment' (EChA, 2011).

6.2 Practical Risk Assessment Processes

Since engineers and scientists performing risk assessments are typically trying to look into the future, uncertainties in their risk assessments are unavoidable – at best, the level of uncertainty may be quantitatively described. Engineers might divide the field of risk assessment into qualitative, semi-quantitative and quantitative approaches. In each case, the work should be based on a structured approach to ensure the assessment is as complete as possible and that the user of the assessment is aware of the ways in which it is incomplete or uncertain.

6.2.1 Qualitative Approaches

Qualitative approaches are basically about identifying risks rather than trying to be too specific about which ones are the worst. Two of the most famous are HAZOP (HAZard and Operability study) and HACCP (Hazard Analysis and Critical Control Points). Both were developed in the 1960s.

HAZOP was developed under the leadership of British chemicals company ICI and became widely used, especially after the fatal Flixborough Disaster of 1974, which drove the increased popularity of hazard management coursework put together by the Institution of Chemical Engineers. The approach is well known, especially in the chemical, pharmaceutical and nuclear industries.

A HAZOP study is typically performed by a small, multidisciplinary group of people – between four and eight. Each person should have a defined role, and the full-time roles are: facilitator, scribe, process designer and operator. Technology specialists, maintenance engineers and others may be drawn in from time to time, making for larger groups, but larger group size slows down progress and is obviously more expensive.

A HAZOP study team will typically look at piping and instrumentation diagrams and/or flow diagrams for a facility and segment it into 'nodes', or sections where each has some identifiable 'design intent'. Then they run through a list of standard *guide words* and identify what could go wrong in each part of the node. The words are deliberately simple so they do not constrain the imagination too much, like: *more, less, no, reverse, early, late* and *instead*. For example, Figure 6.1 shows a very simple piping and instrumentation diagram for a railcar filling station, where methanol is loaded onto train tanker wagons for transport. Table 6.1 lists some things which might come up in a HAZOP study of this node.

As this example indicates, HAZOP studies are a rigorous, methodical way to look for things that could go wrong in a process plant. Once a plant or a plan for a plant has been modified on the basis of a HAZOP study, the team should be gathered again to repeat their process. Work like this relies on the alertness, imagination and technical understanding of the team sitting in the HAZOP meetings to ensure that significant risks are identified, so running a HAZOP study requires risk managers to ensure that the best staff members are put in this role, and that they are allowed the time and low-stress working environment necessary to enable good communication.

	Process			
Table 6.1 Example HAZOP Study Output				
Guide word	Process deviation	Potential cause	Potential consequence	Action required
Less	Less inflow	Drain valve accidentally left open	Soil contamination	Valve should be automatically and manually checked before filling
More	More inflow	Failure of level indicator (LI)	Railcar overpressure and rupture	Install independent high-level alarm
No	No inflow	Inlet valve accidentally left closed	Electric pump motor burnout, fire risk	Install flow monitor on pump line

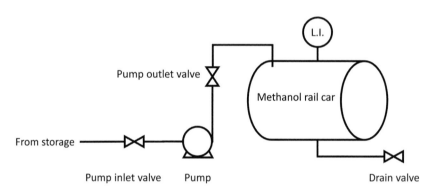

Figure 6.1 Railcar filling station – simple piping and instrumentation diagram (LI = level indicator)

HACCP came about through NASA's need to be certain about providing safe food for astronauts without testing almost all the food. It has become an important approach to risk management in many industries, centred on the food and agricultural sectors.

HACCP has seven important elements, according to the relevant ISO standard (ISO 22000). These are described in Table 6.2.

One can imagine the application of HACCP to the operation of a packaging plant in a fresh food manufacturing operation – Table 6.3 provides some examples of what might be included in HACCP plans for such a facility. None of these elements of HACCP should seem radical or surprising. Nevertheless, there are plenty of examples of situations where a complete approach like this was not fully in place. For example, disaster can strike when a process is monitored, and workers may notice when a critical value is exceeded, but in the heat of the moment, they do not know how to interpret it or what the most appropriate corrective action is. The lead-up to the Fukushima nuclear disaster of 2011 is an example of this.

Table 6.2 Principal Elements of HACCP
1. Perform a hazard analysis – identify the biological, chemical or physical properties of the product that may make it dangerous.
2. Identify critical control points – places in the process where a procedure to reduce the hazard can be applied.
3. Determine critical values associated with each control point – a maximum or minimum cooking temperature, for example.
4. Establish monitoring requirements – what kind of monitoring should be applied and how often?
5. Establish corrective actions – determine what staff should do if a critical value is breached.
6. Identify the test procedures necessary to tell whether the HACCP plan is robust and working. This could involve simulating a breach and observing the response.
7. Establish record-keeping methods – ensure that all of the previous elements are documented, breaches and corrective actions are recorded, and thus a data basis for continuous maintenance of the HACCP plan is created.

Table 6.3 Simple Example Content in an HACCP Plan			
Hazard details	Freezer room air temperature	Food pallet spill	Staff sneezing on packaging line
Critical control point	Electronic thermometer	Visual identification by staff	Shift kick-off meeting
Critical limits	Must not exceed −28°C	No residue on floor	No unwell staff on packaging line
Frequency	Continuous	Hourly inspection on foot	Daily
Corrective action	Activate standby chiller	Mess clean-up kit in central store	Packaging staff sent home or swapped to office role
Responsible person	Shift supervisor	Maintenance supervisor	Shift supervisor
Documentation	Factory data server	Daily process report	Daily process report

6.2.2 Semi-Quantitative Approaches

Let's imagine you have to assess some risks for which the quantitative data are rather weak. One way to do this is to use a risk assessment matrix. This approach has been adopted by a number of guidelines for risk assessment (e.g. NHMRC, NRMMC, 2011). Instead of trying to carry quantitative exposure data all the way through the assessment to a quantitative bottom line, the analyst uses the data at hand to estimate a qualitative descriptor of likelihood like one of those shown in Table 6.4.

The analyst can also select a descriptor for the consequence or impact of a hazardous event (see Table 6.5). As you can see, these tables are not completely qualitative, and can include some degree of numerical information. The level of risk

Table 6.4 Qualitative Descriptors of Likelihood (Adapted from NHMRC, NRMMC, 2011)		
Level	Descriptor	Example description
A	Almost certain	Is expected to occur in most circumstances, between daily and weekly
B	Probable	Will probably occur in most circumstances, weekly
C	Possible	Might occur or should occur at some time, once a month
D	Unlikely	Could occur at some time, every year
E	Rare	May occur only in exceptional circumstances, once a decade

Table 6.5 Qualitative Descriptors of Impact		
Level	Descriptor	Example description
1	Insignificant	Insignificant impact, little disruption to normal operation, low increase in normal operation costs
2	Minor	Minor impact for small population, some manageable operation disruption, some increase in operating costs
3	Moderate	Minor impact for large population, significant modification to normal operation but manageable
4	Major	Major impact for small population, systems significantly compromised and abnormal operation if at all, high level of monitoring required
5	Catastrophic	Major impact for large population, complete failure of systems

is then characterised by using a qualitative risk matrix, like the one shown in Table 6.6.

These three tables were drawn from government guidelines for drinking water quality risk management. They can be adapted to a variety of circumstances and industrial situations. Imagine, for example, their potential use in examining the risk associated with washing an aircraft using industrial detergents. Splashing contact with the eyes of the worker doing the washing is almost certain to happen (A), so one would direct the workplace to use detergents that are not blinding, but merely irritating – minor impact (2). This frequent irritation could be a problem, not the least for worker morale, as indicated by the 'high' rating. If the risk manager chooses a policy of reducing all risks to 'low', this might be achieved by ensuring the use of protective goggles, thus lowering the likelihood to unlikely or rare (E). This qualitative analysis is illustrated in Table 6.7.

6.2.3 Quantitative Risk Assessment

Figure 6.2 shows a generic quantitative human health risk assessment process intended for the assessment of environmental hazards (adapted from DHE, 2002).

Table 6.6 Qualitative Risk Level Matrix					
	Impact				
Likelihood	1	2	3	4	5
A	Moderate	High	Very high	Very high	Very high
B	Moderate	High	High	Very high	Very high
C	Low	Moderate	High	Very high	Very high
D	Low	Low	Moderate	High	Very high
E	Low	Low	Moderate	High	High

Table 6.7 Qualitative Risk Level Matrix for Aircraft Washing Example					
	Impact				
Likelihood	1	2	3	4	5
A	Moderate	High	Very high	Very high	Very high
B	Moderate	High	High	Very high	Very high
C	Low	Moderate	High	Very high	Very high
D	Low	Low	Moderate	High	Very high
E	Low	Low	Moderate	High	High

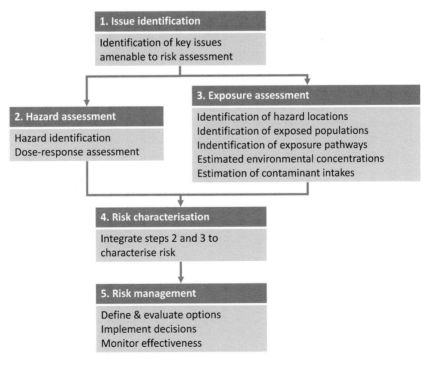

1. Issue identification

Identification of key issues
amenable to risk assessment

2. Hazard assessment

Hazard identification
Dose-response assessment

3. Exposure assessment

Identification of hazard locations
Identification of exposed populations
Indentification of exposure pathways
Estimated environmental concentrations
Estimation of contaminant intakes

4. Risk characterisation

Integrate steps 2 and 3 to
characterise risk

5. Risk management

Define & evaluate options
Implement decisions
Monitor effectiveness

Figure 6.2 Generic quantitative risk assessment process

The remainder of this chapter takes you through each of the five steps shown in the figure, describing their qualitative and quantitative aspects. You should also consider the contents of other chapters, particularly Chapter 5 in relation to step 2 and Chapter 4 in relation to step 3.

6.2.3.1 Issue Identification

This first step in risk assessment provides the basis for hazard assessment and exposure assessment. It involves answering a few questions listed in the discussion here. Documenting this step is an important element in defining and scoping any risk assessment – providing transparency and protecting the analyst from accusations of bias. If you are up-front with which risks you intend to evaluate, it is hard to accuse you of attempting to mislead a decision-maker. The context of a risk assessment is relevant to note here. Some of the questions that are worth getting answers to are: *who raised the initial concern about a potential risk?* Was it a scientist, a politician, a member of the public? Noting this information is relevant so the reader can understand who the intended audience is and what kind of information and level of detail will be necessary to satisfy the audience's need for information. Furthermore, the analyst ought to engage with such persons as need be in order to obtain further detail on the kinds of hazards they perceive and the circumstances of exposure they consider to be of most concern. *What were the circumstances that triggered concern about a risk?* This question concerns the hazards and exposure routes connected to the risk. Of course risk assessment professionals may want to add additional hazards and exposure routes to the scope of a risk assessment based on their own initial perceptions and the perspectives of other experts. It may be that for practical or toxicological reasons, the analyst may need to prioritise one hazard over another in the analysis. Alternatively, it may be that the hazards interact in some way that necessitates their parallel evaluation. *What is the scope of the issue?* A risk assessment might be performed to evaluate the potential human health impacts on people working in a particular laboratory in a particular university, or the scope may be citywide or national. In each case, the hazard assessment and exposure assessment will have to be adjusted to properly reflect the scope of the issue. It has also to be asked initially: *is it feasible to perform a risk assessment for the concerns raised?* It may be that the data necessary for a risk assessment do not exist, and decision-makers need to instead be lobbied to provide more resources to generate the data – or that while a quantitative risk assessment was sought or is considered ultimately necessary, a qualitative risk assessment is the only feasible assessment at the current time. Analysts should make these points early in their assessment work if they are appropriate to the particular case.

6.2.3.2 Hazard Assessment

The second step in Figure 6.2 is hazard assessment, and its first element is hazard identification. At this point, we need to be clear about the difference between a hazard and a risk, since in everyday language they may seem interchangeable. *Risk* was defined in Section 6.1. According to the Cambridge dictionary, a *hazard* is 'something that is dangerous and likely to cause damage'. The word *stressor* is also used for this. Our central examples in this book relate to chemical hazards, but hazards of concern to the engineering professional are not limited to chemicals. Human and ecological health can be affected by various forms of energy, including light, heat and noise, not to mention radioactivity. Pathogens are also a key concern for engineering professionals in the food industry and the waste management sector. Pathogens are of particular interest on account of their potential for seeding and regrowth in otherwise uncontaminated media.

Let's say you ask the question: *which chemicals are of concern in the proposed product?* One can begin to answer this example question by looking at the ingredients of the product, including potential contaminants, and the potential reaction products that arise when they are combined. One should also consider the possibility that chemical emissions from the original product (primary pollutants – see Section 3.1) may not be the most critical, but that the products of microbial or chemical break-down of the initial emissions (secondary pollutants) may be more toxic. It could also be that during waste management, the potential toxicity of a product increases, for example when chromated copper arsenate is used as a preservative for outdoor wood products, the potential for leaching is modest, but if the wood is mistaken for any other wood and burnt in the domestic environment, the smoke and the ash are of serious concern.

Sometimes it is necessary to invest in analytical chemistry to identify whether hazardous contaminants may be present in a certain place, for example in wastewater or in land that has been used for industrial purposes. Three principal approaches are used here: *in vivo, in vitro* and chemical analysis. *In vivo* ('within the living') refers to animal testing in a living organism, like a laboratory rat. *In vitro* ('in glass') tests use instead cultured cells, microorganisms or biological molecules in a laboratory vessel (not necessarily a test tube, though these are sometimes called 'test-tube experiments'). They are a popular alternative to *in vivo* studies, being quicker and less ethically challenged than animal testing. Analytical chemistry covers a wide range of approaches from classical wet chemistry (e.g. titration) to modern spectroscopic, electrochemical and thermal methods of identifying chemicals in a sample. Obviously these three testing approaches provide different kinds of information and have different benefits and drawbacks – some are summarised in Table 6.8.

As indicated in Table 6.8, there are advantages in moving beyond analytical chemistry to identify hazards in a real sample of water, air or soil taken from the environment. This is particularly true for the problem of discovering whether there may be additive or synergistic effects of several contaminants, and whether the overall sample is hazardous. Different analytical chemistry methods are good for finding different types of contaminants, so the chemist will only find the chemicals her methods can find. In other words: she can only find the chemicals she decides to search for.

Another key question here is: *what type of health or environmental effects may be caused by the hazard?* If the analyst is lucky, retrospective data on relevant hazards may be available in literature. There is also a wide range of field studies which may be relevant for hazard assessment, for example the items shown in Table 6.9. When published information is inadequate, laboratory *in vitro* and *in vivo* methods can be

Table 6.8 Approaches to Hazard Identification

Test approach	Analytical chemistry	In vitro	In vivo
Is it relatively fast?	Yes	Yes	No
Will it identify a particular chemical in a mixed sample?	Yes	Depends on how specific the method is	No
Will it identify mixture toxicity?	No	Maybe	Yes
Will it indicate the presence of 'unknowns'?	No	Yes	Yes

Table 6.9 Field Data Sources for Hazard Identification

Data source	Example data
Environmental monitoring	Food, air, soil quality monitoring results
Biological monitoring	Lead levels in the blood of school children
Passive disease surveillance	Liver cancer rates reported by hospitals
Active health monitoring	Lung function test results for employees at risk of environmentally induced asthma
Directed epidemiological studies	Disease rates identified by medical outreach programmes to particular populations

used to answer this question. For hazardous chemicals, there is also a range of *in silico* (computer-based) approaches based on structural analysis of the hazardous chemical – the same starting point Meylan and Howard (1995) used to estimate the octanol-water partition coefficient (described in Chapter 4).

The second key element of hazard assessment is the establishment of dose-response relationships. As described in Chapter 5, when an organism is exposed to a certain environmental concentration of a contaminant, whether pathological or chemical, there are various defence mechanisms concerned with limiting absorption. These may operate at the contaminant's target organ or elsewhere in the organism, encouraging elimination from the organism, possibly via metabolism of the contaminant. These mechanisms mean that dose-response relationships are not typically linear. An example of such a nonlinear relationship is shown in Figure 6.3. The sigmoid character of this example, in particular the low slope in the dose-response curve between zero and 5 mg/(kg.day), may be caused by the organism's ability to eliminate the contaminant from its body. As the dose increases above this level, the organism's defences are overwhelmed and small increases in dosage create a larger response (higher fatality rate among test organisms). The curve may flatten out towards the top right because of the survival of few abnormally hardy individual organisms in a test population, who are especially resistant to the hazard.

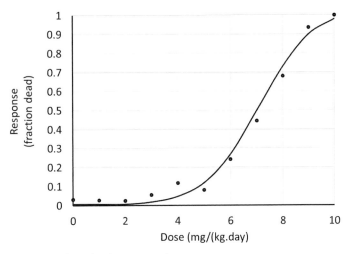

Figure 6.3 Example dose-response relationship for a non-carcinogen

In microbial risk assessment, the dose may be in terms of the number of pathogenic organisms absorbed rather than the mass of a chemical, as shown in the figure, and the effect on the y-axis may be 'probability of infection'.

Looking at a figure like Figure 6.3, the analyst can visually identify a key descriptor of hazard: the LD_{50}. This is the dose that is lethal for 50% of the organisms in a study, in this case about 7 mg/(kg.day) (in this case, daily dose rate versus animal body mass). A number of risk assessment activities are based on comparing the LD_{50} of different contaminants, in the context of other parameters like exposure rates. In the case that the y-axis of the figure does not describe mortality but some other endpoint like the onset of a health condition, or in the case of the therapeutic use of pharmaceuticals, some desirable endpoint like healing, we may instead speak of an ED_{50} where E stands for *effective* rather than L for *lethal*.

Another key indicator of toxicity is the NOAEL or *no observable adverse effect level*, which is the lowest dose at which no response is observed. In human health risk assessment, given a NOAEL (for example, 2 mg/[kg.day]), a risk analyst may calculate a reference dose (RfD). The USEPA defines this as '*an estimate, with uncertainty spanning perhaps an order of magnitude, of a daily oral exposure to the human population (including sensitive subgroups) that is likely to be without an appreciable risk of deleterious effects during a lifetime*' (USEPA, 2018). Clearly, this has to be lower than the NOAEL in order to deal with the uncertainty in the dose-response data, the variation between experimental conditions and the variability among people. So the RfD is calculated by dividing the NOAEL by an *uncertainty factor* (UF). The USEPA's IRIS database (www.epa.gov/iris) is a useful place to find NOAELs. For example, it suggests that acenaphthene, a combustion by-product and a chemical used in textile and paper production, has a NOAEL of 175 mg/(kg.day) based on hepatotoxicity observed in experiments on mice. The USEPA applies a factor of 10 to account for the variation between rodents and people, 10 for variation among people, another 10 to take into account that data from a subchronic (short-term) experiment were used to derive a chronic (long-term) RfD and another 3 to account

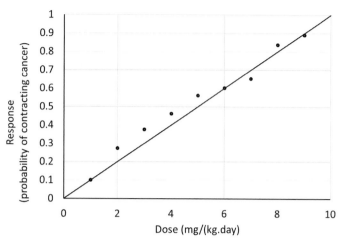

Figure 6.4 Dose-response relationship for a carcinogen

for the fact that only one species was tested and that reproductive/developmental data are missing. So the UF is 3,000 and the RfD is 0.06 mg/(kg.day).

The sigmoidal curve is common in chemical risk assessment, but a variety of curve shapes other than this are possible. A key contrast is drawn between the dose-response curve for carcinogens and non-carcinogens. By contrast, carcinogens are not considered to have a safe dose – a single molecule could give you cancer, but only if you are extremely unlucky. Figure 6.3 is a possible curve for a non-carcinogen – compare this with Figure 6.4.

In this figure, the slope of the curve near the intercept for a dose of 1 mg/(kg.day) is called the 'slope factor' with units of (kg.day)/mg. This value is compared with estimated doses in risk assessments to provide an upper estimate for carcinogenic risk.

6.2.3.3 Exposure Assessment

If the aim of hazard assessment is to answer a question like 'How bad is it if I eat some acenaphthene?' then the purpose of the exposure assessment is to answer 'If there are acenaphthene emissions from a factory near a lake where I catch fish, how much am I going to eat?' More generally, this means constructing a model of the pathways from the hazard source to the exposed individual (person, organism or ecosystem) and evaluating the potential for the hazard to be attenuated by dilution or barriers. As shown in Figure 6.5, there may be multiple pathways for exposure to a particular contaminant. In some cases, the hazard may become more significant from one end of the pathway to the other via some kind of reaction (for chemicals) or by regrowth (for pathogens).

Chapter 4 of this book provides more information about basic approaches for modelling the environmental transport and fate of contaminants. In addition to finding out what the concentration of a contaminant is likely to be in some compartment of the environment (a lake, an indoor airspace, etc.) in which an organism is exposed, the analyst must consider how often the organism will be

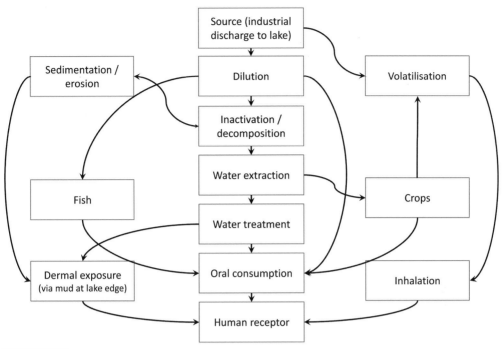

Figure 6.5   Possible exposure pathway diagram for a chemical discharged into a lake

exposed and to how much of the contaminated compartment. Consider as a specific example of this a risk assessment to be performed to evaluate human health impacts associated with living in a contaminated urban residential environment. The USEPA *Exposure Factors Handbook* (USEPA, 2011) provides a standard intake rate of 2 L of drinking water, 20 m^3 of air and 100 mg soil per day. You may wonder: do people really eat soil? For adults, much of this is airborne dust delivered via inhalation and the body's mechanism for keeping airways clean – flushing to the stomach (see Chapter 5 on toxicology). For children, the EPA recommends using half the drinking rate, but, for small children (under seven years old), twice the soil consumption rate! This is because small children often love to play in the dirt. For lifetime exposure calculations, the *Exposure Factors Handbook* suggests using an exposure frequency of 350 days per year over 30 years for drinking water and air inhalation at home. This is based on the assumption that people are typically away from home on holiday for a fortnight per year. The *Exposure Factors Handbook* also provides guidance on intake rates, exposure frequencies and exposure durations for workplace environments.

6.2.3.4 Risk Characterisation

This fourth step in risk assessment is where we bring together the outcomes of the hazard assessment and exposure assessment, and use them to answer the central question of how likely it is that the hazard will cause unwanted outcomes. There are various ways to do this. Assuming you have a reasonable, quantitative exposure model for a contaminant, and a trustworthy dose-response curve, then quantitative approaches to characterising risk are available. In chemical risk assessment, we distinguish between approaches for non-carcinogens and carcinogens.

Non-Carcinogens

The risk of health effects caused by non-carcinogens is typically described using a hazard quotient (HQ). This is simply the ratio of the reference dose and the actual dose of a contaminant:

$$RQ = \frac{dose}{RfD} \tag{6.1}$$

When $RQ \geq 1$, we say that there is a possibility of health effects. Note that since the RfD is based on a NOAEL and an uncertainty factor rather than any knowledge of the slope of a curve around the EC_{50}, it would be wrong to say that when $RQ = 2$, the risk is twice that when $RQ = 1$. On the other hand, if $RQ < 1$, there is no appreciable risk of adverse effects.

You can reorganise Equation 6.1 to use it to determine the acceptable concentration of a contaminant under certain exposure conditions (e.g. an individual of a certain mass, with the intake rate appropriate for their age), as the example in Equation 6.2 shows:

$$conc_{acc} = \frac{HQ \times RfD \left[\dfrac{mg}{kg.day}\right] \times \text{bodyweight}[kg]}{\text{intake rate} \left[\dfrac{L}{day}\right]} \tag{6.2}$$

Carcinogens

The basic equation for the assessment of the lifetime risk of cancer due to a hazardous chemical is given by Equation 6.3:

$$risk = dose \left[\frac{mg}{kg.day}\right] \times slope\ factor \left[\frac{kg.day}{mg}\right] \times \frac{exposure\ time[years]}{lifetime\ length[years]} \tag{6.3}$$

Notice that this takes the weight of the exposed people into account – this is why kg is included in the denominator of the dose term. The *lifetime length* (also called *averaging time*) is usually assumed to be 70 years of 365 days each, but could be a shorter period if the analyst is not calculating lifetime risk. *Exposure time* is the exposure frequency (e.g. days per year) multiplied by the exposure duration (in years).

You can flip this equation to calculate the acceptable concentration of a carcinogen in a compartment. In Equation 6.4, this is shown in the more general form for an exposure averaging time (rather than a lifetime in particular):

$$conc_{acc} = \frac{acceptable\ risk \times bodyweight[kg] \times averaging\ time[days]}{slope\ factor \left[\frac{kg.day}{mg}\right] \times intake\ rate \left[\frac{L}{day}\right] \times exposure\ time[days]} \tag{6.4}$$

In this equation, *acceptable risk* is a dimensionless variable, like 10^{-5} for one cancer occurrence per 100,000 people.

Pathogens

In quantitative microbial risk assessment, analysts calculate the risk of infection. There are several possible curves, but typically an exponential or beta-Poisson function is used (Haas, 1996):

$$P = 1 - e^{-\left(\frac{N}{k}\right)} \tag{6.5}$$

$$P = 1 - \left(1 - \frac{N}{\beta}\right)^{-\alpha} \tag{6.6}$$

$$P = 1 - \left[1 + \frac{N}{N_{50}}\left(2^{\frac{1}{\alpha}} - 1\right)\right]^{-\alpha} \tag{6.7}$$

The second two of these are forms of the beta-Poisson function. In these equations, P is the probability of infection or illness, N is the number of ingested pathogens (the dose), k is the average number of pathogens that must be ingested to initiate infection (the reciprocal of the individual pathogen survival probability), N_{50} is the average dose infecting half the population and α and β are two dose-response parameters. Numerical values for these parameters for specific microbes can be found in sources like USDA (2012).

Burden of Disease Indicators

A risk assessment may be concerned with a number of different unwanted human health outcomes. Several methods have been proposed as ways to aggregate the consequences of various accidents or illnesses. Two of the most widely used ones are Quality Adjusted Life Years (QALYs) and the alternative Disability Affected Life Years (DALYs).

QALYs are positive indicators: 1 QALY represents a person who is perfectly healthy for a year. The influence of various illnesses has been turned into tables of QALY weights, for example by asking people if they would be prepared to exchange 10 years of living with a certain illness, with a smaller number of years of perfect health. The smaller the number, the bigger the impact of that illness is.

DALYs are more popular in risk assessment and life cycle assessment. They are negative indicators: the sum of years lived with a disability (YLD) and years of life lost (YLL) within a population:

$$DALY = YLL + YLD \tag{6.8}$$

So if you are expected to die five years early on account of a risk, or five people are expected to die a year early, those outcomes are in both cases five DALYs. To calculate YLD, we use the formula:

$$YLD = I \times DW \times L \tag{6.9}$$

Table 6.10 Some DALY Weights (WHO, 2013)	
Health outcome	Disability weight
Mild diarrhoea	0.061
Blindness	0.195
Burns to 20–60% of body	0.438
Metastatic cancer	0.484

where I is the number of cases in a population of people, DW is a disability weight and L is the average length of the health condition until health is restored, or death. The World Health Organization has adopted the use of DALYs and produces an authoritative table of DW for years lived with different conditions. Some examples of these which would be relevant for chemical and pathogen exposure are shown in Table 6.10.

6.3 Communicating Risks and Risk Management

Once the risk under consideration in the previous steps of the process has been characterised in step 4, the decision-maker has to ask: *is the risk acceptable or not?* This is a very interesting question because individuals tend to perceive the importance of different risks in proportion to how unusual they seem or how much media attention they get, rather than how likely they are to cause a major impact. Researchers explain differences in risk perception in various ways. For example, psychologist Paul Slovic identified 10 factors that influence how we perceive risks (in Hillson, 2009): *dread* (we imagine terrible, awful outcomes; we tend to exaggerate), *control* (which is even often only an illusion), *nature vs man-made* (natural disasters often seem less risky than human-created ones), *choice* (if we have a choice, consequences seem less risky), *children* (we are programmed to care for children), *novelty* (risks that we have not encountered before generate more concern), *publicity* (public attention makes a risk appear more significant than it actually is), *propinquity* (if I am the subject of risk, I am likely to assess the risk as being higher than if I am a bystander), *risk-benefit trade-off* (if there are opportunities as well as risk, this can make the actual risk appear to be less than it actually is) and *trust* (risk will be significantly affected by the extent to which we trust other parties involved in managing the risks). Some of the factors reducing our perception of risk are exemplified in Table 6.11. In summary, many factors may cause two individuals to respond to a risk differently, even when they are presented the same information.

Sometimes risk perceptions are a matter of national culture. We have heard Europeans expressing travel-stopping fear of Australia's poisonous spiders and snakes, while many Australians express more concern about the risks that reckless taxi drivers pose to passengers and other drivers in many countries, including their own. There is

Table 6.11 Risk Attributes That May Lower or Raise Risk Perception		
Characteristic	Example activity	Counter-example
familiar	eating unhealthy food	exposure to exotic snakes
controllable	driving your own car	sitting in an aircraft
natural	solar UV radiation	working with industrial radiation
voluntary	recreational skiing	military conscription

some statistical support for the latter risk perception – in Australia in 2015, a total of 712 people died while travelling in cars, but only three died from venomous bites – more died after being struck by lightning (i.e. 5 – ABS, 2016). To aid comparison between countries and regions with different total populations, analysts like to convert data of this kind into mortality per 100,000 people, sometimes after adjustment for the local age distribution (AIHW, 2011). Those Australian car occupants then represent a death rate of 3 per 100,000 – total transport accidents (including other vehicles and pedestrians) occurred at a rate of 6 per 100,000. This is similar to the mortality data from the United States, where there is less than a 1 in 10 million chance of dying on account of an attack by an animal, compared to a 13 in 100,000 chance of dying in a traffic accident. The risk of death by gunfire in the United States (excluding suicides) is about 4 in 100,000 (CDC, 2015), which is extremely high compared to other Anglo-Saxon countries with stronger gun laws – that rate is 0.2 per 100,000 in Australia (ABS, 2016) and 0.08 in the UK (WHO, 2016). All of these countries have relatively democratic legislatures, which suggests that people in some countries are more concerned about managing certain kinds of risk than other people. When government regulatory agencies take a stance on the scale of an acceptable risk, they typically use a threshold that ranges from 1 in 10,000 to 1 in 1 million, with most policies set at either 1 in 100,000 or 1 in 1 million (Mihelcic & Zimmerman, 2010).

Exactly how to communicate the risk to decision-makers is a subject of some variation and debate. For example, Masters and Ela (2008) suggest several different ways to describe the risk associated with smoking. One approach is to describe the annual risk experienced by a person exposed to the particular hazard – in this example, there is a 300 in 100,000 chance that if you are a smoker, you will die in a year. This can be compared with other voluntary risks (e.g. an annual fatality risk of 80 in 100,000 hang-gliders) or perhaps some of the involuntary risk data in the previous paragraph. Alternatively, one might compare activities on the basis of a standard risk. For example: smoking 1.4 cigarettes increases your mortality risk by 1 in 1 million, like driving a car 482 km.

A key challenge in communicating risk is the need to express the level of uncertainty associated with what is by nature a model estimate. On the one hand, explaining that there is uncertainty may discourage recipients of risk assessments from treating identified risks as real. On the other hand, expressing the uncertainty in risk estimates is also a necessary part of establishing trust between risk assessors and the audience for their work. Another key aspect of

Table 6.12 Global Harmonised System Hazard Pictograms

Explosive		Harmful (irritant, narcotic effects)	
Flammable		Health hazard (sensitising, mutagenic carcinogenic, specific organ toxic)	
Oxidising		Corrosive	
Compressed gas		Environmental hazard	
Toxic (acute toxicity)			

risk communication is ensuring that the risk assessment is easily understood. An example of an effort in this direction is reflected in the Globally Harmonised System (GHS) hazard pictograms, shown in Table 6.12. These have been adopted by the European Union.

6.3.1 Exposure Limits

Various national regulatory authorities have created exposure limits to ensure that employers have practical, risk-assessment-based standards for ensuring the occupational health of their employees. By nature, the standards simplify the range of possible exposure scenarios, but they are much easier to communicate and measure than more complex alternatives. Examples of these standardised exposure limits are those set by the Swedish Work Environment Authority (Arbetsmiljöverket) (AFS, 2015). Exposure limits are set at two levels:

- eight-hour, time-weighted average (TWA) (or, in Swedish, *nivågränsvärde*); and
- short-term exposure limit (*korttidsgränsvärde*), which is typically for a reference period of 15 minutes, but, in a few cases, for five minutes.

Chemical	Eight-hour TWA (mg/m³)	Short-term exposure limit (mg/m³)	Notes
Acetaldehyde	45	90	C, V
Acetone	600	1,200	V
Acetonitrile	50	100	H, V
Acrylamide	0.03	0.1	C, H, V, M
Ammonia	14	36	
Aniline	4	8	C, H, V
Turpentine	150	300	H, S, V
Tetraethyl lead	0.05	0.2	H, R, V

Table 6.13 Some Exposure Limits (AFS, 2015)

C: Carcinogen
H: Dermal uptake route
M: Medically regulated substance, use may require a doctor's authority
R: Toxic to reproduction
S: Sensitising substance
V: The indicative short-term value is also a maximum threshold limit.

Feasibility analysis
• Technical feasibility
• Cost-benefit analysis
• Social factors
• Ethical factors
• Legislative factors

Risk reduction
• Classification and labelling
• Safety standards
• Technical measures
• Organisational measures
• Personal protection

Monitoring and review
• Hazard monitoring
• Audit of risk management system
• Risk (re)assessment

Figure 6.6   Generalised approach to risk management

Another possibility used previously in Sweden and elsewhere is a 'threshold limit value' (*takgränsvärde*), which is a maximum for any instant in time. Since these exposure limits are intended to place an upper bound on how much of a contaminant a person may absorb, the longer the exposure period, the lower the limit (see examples in Table 6.13).

6.3.2 Risk Management

In a sense, risk management is like many other kinds of management modelled on the PDCA cycle described in Chapter 8, except in this case, the key performance indicator is risk rather than profitability, product quality or greenhouse gas emissions. So risk management is an iterative process (see Figure 6.6). If the analyst has characterised and assessed the risk associated with a product, process or course of action (right-hand side of the figure), the decision-maker may decide that it is appropriate to intervene and suggest some alternative management actions to reduce risk. These actions will have to be assessed for their technical feasibility, cost and consistency with any relevant legislation in the region where they are to be implemented. There may also be social or ethical factors that make them problematic in a particular context. For example, a policy of excluding disabled people from hazardous workplaces because of a need to make emergency evacuations fast would be considered discriminatory by many people. Actions to reduce risk may be about eliminating the risk entirely (e.g. eliminating a hazardous chemical from a process) or ensuring staff are aware (risk labelling) and protected (personal protective gear). Once such risk-reduction actions have been implemented, it is important not to merely assume the risk has been managed, but to monitor the hazard in question (e.g. air quality inside a factory) and regularly audit the risk management system to check if it is being properly implemented. This will hopefully indicate there is no need for further action, but if not, the cycle must begin again.

6.3.3 Levels of Containment

When it comes to the management of risks, it makes sense to have a scale of responses to the risk, rather than using every available means to manage everything from mildly hazardous agents (which would be expensive 'overkill') to those which are extremely hazardous.

Table 6.14 Biosafety Levels (BSL) and Associated Measures				
Biosafety level	BSL 1	BSL 2	BSL 3	BSL 4
Example relevant hazard	Non-pathogenic *E.coli*	Pathogenic *E.coli*, hepatitis, *Salmonella*, HIV	Tuberculosis, Yellow fever, West Nile virus	Ebola virus, Hendra virus
Example safety measure (cumulative left to right)	Mechanical pipetting, eye protection, gloves, lab coat	Biosafety cabinet, disposable syringe units, puncture-resistant waste containers	Sealed windows, air filter within lab, lab coats with rear access only	Positive pressure suit for staff, airlock entry to lab
Access (cumulative left to right)	General staff	No immuno-compromised staff	All staff immunised	All persons entering and exiting lab are recorded
Decontamination (cumulative left to right)	Disinfectant	Autoclave or other method	Autoclave	Exiting air must be sterilised

An example of such a scale is the classification of laboratories for handling infectious hazardous agents. The US Centers for Disease Control has set up a scale of four biosafety levels (BSL) to regulate laboratories and manage microbial risk (USDHHS, 2010). These state that when certain hazardous agents are going to be used in research, the laboratory must meet certain requirements. Examples of the kinds of hazardous agents and the requirements for their management are shown in Table 6.14. The European Union implemented the use of the same levels via a Council Directive (90/679/EEC).

Table 6.14 is also an example of a *multiple-barrier risk management approach* – under BSL 4, the safety measures in BSL 1–3 are not abandoned, but at each BSL there are additional barriers to infection to reduce risk. This is not a complete list of requirements – the interested reader should go to the sources cited earlier in this chapter for more information. This same approach is embodied in centralised drinking water treatment systems. When source water does not come from nature reserves and may contain human pathogens, a series of treatment steps may be implemented, for example:

1. Removal of suspended solids (coagulation, flocculation, sedimentation);
2. Filtration (for example, through a sand filter);
3. Chlorine disinfection (addition of Cl_2 gas or hypochlorite);
4. UV disinfection (exposure to intense light); and
5. Booster chlorination (addition of hypochlorite to water mains).

Each of these 'barriers' to infection has some capacity to eliminate pathogens, and while any one of these barriers may not be adequate for the production of safe drinking water, the presence of multiple barriers creates redundancy in the treatment system design so that if one barrier fails, the likelihood of a severe health outcome is reduced.

6.4 Criticism of Risk Assessment

Like any data-intensive information process, risk assessment is subject to problems such as incomplete or obsolete data, oversimplification of models and the like. In addition, in recent times, a 'social constructivist' critique of risk assessment concepts has developed. More specifically, it proposes that risk assessments:

• Leave out diverse risks that normal people perceive;
• Involve averaging over populations, when individual sensitivities or values may exhibit wide variation;
• Cannot be seen as purely scientific products, since normative values are part of the characterisation and interpretation of risk; and
• Combine probability and severity in equal measure, while in practice, people are more concerned about severity than probability.

Some representatives of this school of thought go as far as saying that all risk assessments are just social constructs, even quantitative human health risk assessments. These critics would say that the measures employed (financial losses, deaths) also have no objective value, but are based on subjective social values (Nilsson, 2003). On the other hand, this kind of criticism could be directed at just about any kind of information produced to inform decision-makers and does not make human health or ecological risk assessment a less respectable or useful activity than the fundamental financial analyses of businesses which are performed by banks (investment risk assessment), or the work of triage nurses who are critical to the management of healthcare facilities. (They have to assess whether a patient should receive urgent help, or can wait, or will die in either case – a form of rapid health risk assessment.) So despite the theoretical or practical criticisms, risk assessment of one kind or another is a necessary skill for many professions and will not be disappearing anytime soon.

6.5 Review Questions

1. Your friend gets a job in a small fast food restaurant, but is worried about staff getting hurt there. Imagine the place and use the description of the HAZOP method in this chapter to identify the relevant risks and control strategies.
2. Using the results of the previous question, perform a qualitative risk assessment on the restaurant, with and without the control strategies.
3. What is a NOAEL? What is an RfD? And how do you convert the former to the latter?
4. Your city lets a contractor dump a persistent organic compound in a simple landfill a kilometre from your house. Draw an exposure pathway diagram explaining how you could be affected.
5. Explain what DALYs are used for and how to calculate them.
6. For a change of pace, compete with a friend in a game of GHS Pictogram Awareness Bingo! How many different GHS symbols can you find on products in your home? Are there more in your friend's home? Which extra pictogram(s) did the winner find? What kind of consumer product(s) needed them?
7. This chapter described the idea of multiple-barrier risk management in relation to water treatment and biohazard labs. Find a description of how high-strength nuclear waste is managed. How many barriers are in place in this case? More than for water treatment?

7

Environmental Assessment of Products and Processes

This chapter is intended to give you a practical overview of some of the key environmental assessment tools in use by applied scientists and engineers working on product and process design, policy development and other fields. A key benefit of having applied scientists with environmental assessment skills as members of a multidisciplinary team is their ability to evaluate options quantitatively and help integrate multiple perspectives (e.g. environmental and financial). Some ways to bring different perspectives together are discussed in more detail in Chapter 9. This chapter focusses on the evaluation tools that contribute some of the relevant perspectives.

7.1 The Life Cycle Perspective

One of the key developments in environmental management and design over the past 25 years has been the increasing emphasis on obtaining a life cycle perspective. The previous generation's paradigm had a focus on point source emissions, embodied in regulatory structures focussed on recipients like airsheds and rivers (e.g. US Clean Air Act, 1970; Australian Clean Air Act, 1961; West German Federal Control of Pollution Act, 1974; EC Directive 70/220/EEC, 1970). Such structures have been refined and developed (see Chapter 8) and still play a critical role in the management of point sources. On the other hand, it has been recognised that once point sources are under control, diffuse emissions need to be managed, and that they are intrinsically much harder to tackle on account of their large number, their wide distribution and, in many cases, their mobility. For example, it has been estimated that the majority of nitrogen pollution in the north-eastern United States comes via atmospheric deposition of oxides of nitrogen, which are produced in combustion processes such as the burning of petrol in cars (see Figure 7.1). So preventing eutrophication in estuaries is a complicated problem of vehicle and transport system design, rather than just a matter of controlling a few wastewater treatment plants.

Tackling diffuse emissions requires the analyst to see many different parts of the *technical system* in order to avoid problem shifting between different parts in the life cycle. For example, it is feasible to design a polyethylene tetraphthalate (PET) bottle recycling system with the aim of reducing resource use and greenhouse gas emissions, but the degree of centralisation will determine whether it is environmentally preferable or not. At some point, the extra distance that waste bottles have to be trucked to deliver them to a recycling facility will produce more greenhouse gases than the avoided production of new PET will prevent. As a starting point, the environmental analyst in a multidisciplinary team needs to be thinking holistically about the system, including the production of raw materials, manufacturing of products, their use and subsequent disposal, that is, the whole product life cycle. This holistic mental activity is called 'life cycle thinking'.

In addition to requiring a more holistic perspective of the technical system, modern analysts also require a wider perspective of the *environmental system* than their predecessors. Environmental problems are becoming increasingly 'wicked', in other words, subject to more complex interdependencies (Churchman, 1967 – see also Section 1.2). For example, engineers may previously have regarded urban wastewater treatment as primarily a question of balancing a community's willingness to pay for water purification against the environmental value of a local waterway. Nowadays, these two factors have to be balanced against the greenhouse gas emissions likely to arise from using additional energy to treat the wastewater. So the necessary perspective of the environmental issues is wider. Moreover, in some locations, future wastewater discharge limits may have to be cut on account of a reduction in rainfall brought on by climate change. So the analysis of environmental interventions needs to be more forward-looking than previously.

In product and process design, taking a holistic perspective of both the relevant *technical* and *environmental* systems is now necessary. Environmental systems analysis (ESA) is a group of systematic approaches for describing human impacts on the environment, which is often applied to products, projects, organisations and policies. There are many ways to perform ESA, which includes the use of tools like material flow analysis (MFA), Material Input Per Service (MIPS), life cycle assessment (LCA) and others described in this chapter.

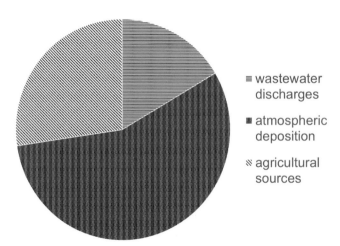

Figure 7.1   Sources of nitrogen pollution in rivers of the north-eastern United States (data from Jaworski et al., 1997)

7.2 Life Cycle Assessment

Though debatable, there are good reasons to claim that LCA is an engineer's central environmental assessment tool, since it is widely used – when we catalogued its applications back in 2009, we found LCA was already being used in 25 different industries around the world (Peters, 2009), and growth in the use of LCA is exponential (Peters et al., 2015). While biologists, social scientists and others are likely to see different environmental assessment tools as more central to their careers, the quantitative character of LCA has placed it predominantly in the engineering profession. Nevertheless, the reason we decided to focus on LCA in this chapter is because it provides a useful general structure for the evaluation of products and systems, and if you have a reasonable understanding of LCA, you may be better able to understand other information tools, such as MFA and input-output analysis (IOA). It is 'central' in the sense that it is possible to make useful comparisons with a wide range of other tools that share some of its elements. So the approach we have taken in this chapter is to firstly examine LCA in some depth, then use it as a way to understand other tools in Section 7.3, along with the strengths and weaknesses of these tools.

7.2.1 History and Framework

The first published LCA studies emerged in the 1970s, driven initially by private companies designing beverage packaging, and then by Western governments worried about the OPEC oil supply crisis. This proved to be something of a false dawn for LCA, but public and private interest in the method grew significantly in the early 1990s, driven by increasing concern about the growth in municipal waste and the limited volume of landfill space available. This was especially apparent after the reunification of Germany in 1990. The Federal German Republic ('West Germany') had exported about 5 million tonnes of waste, including hazardous waste, to the German Democratic Republic ('East Germany'). These deals had earned East Germany useful convertible currency and were good value for West Germany, but they resulted in the deposition of this waste in many poorly managed landfills (Deubzer, 2011). With the reunification, West Germany inherited the problem it had exported. The need to manage solid waste came into sharp focus for the new national government as it was forced to spend billions of Deutschmarks on remediation of East German landfills, and large sums were spent on LCA to analyse alternative waste management scenarios.

The Swiss, Swedish, Danish and Netherlands governments also made large investments in LCA in the early 1990s to inform waste management strategies, and some of the relevant academics published the first comprehensive guide for LCA practitioners, 'Environmental Life Cycle Assessment of Products' (Heijungs et al., 1992). This 'CML Guide' and its later update (Guinée, 2002) spread the LCA methodology to a broader audience and laid the groundwork for global standards published by the International Standardization Organization in the late 1990s. These have been subsequently revised as shown in Table 7.1.

For readers interested in more detail, Baumann and Tillman (2004) provide one of the most extensive and accessible histories of LCA and of the original ISO standards.

Table 7.1 Key ISO Standards for LCA			
Standard	Name	Released	Status
ISO14040	Life cycle assessment – principles and framework	1997	Revised version released 2006
ISO14041	Life cycle assessment – goal and scope definition and inventory analysis	1998	Transferred to ISO14044
ISO14042	Life cycle assessment – life cycle impact assessment	2000	Transferred to ISO14044
ISO14043	Life cycle assessment – life cycle interpretation	2000	Transferred to ISO14044
ISO14044	Life cycle assessment – requirements and guidelines	2006	Current

Briefly, several other international standards of relevance to the life cycle analyst have subsequently been published. Some of these are focussed on particular aspects of LCA like data documentation formats (ISO14048), the use of LCA in product environmental labels (ISO14025) or the calculation of particular life cycle indicators, like carbon footprints (i.e. greenhouse gas emissions) (PAS2050, ISO/TS14067). Most of them can be traced back to ISO14040.

The key elements of ISO14040 are shown in Figure 7.2. Each of these elements is described in turn in subsequent sections of this chapter. While these elements are often called steps or stages in the performance of an LCA, and they are typically performed in the order shown in this book, the double arrows in the figure exist to indicate that there is a high degree of interdependence within the LCA process. So while analysts typically start with the goal and scope definition and write their reports following the figure in an anticlockwise direction (referring to Figure 7.2), robust LCA studies are typically the result of an iterative approach to each element and the process as a whole. For example, it is often only by performing an inventory analysis that an analyst can identify problems with the goal and scope definition.

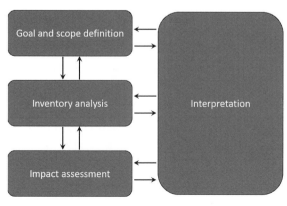

Figure 7.2   Key elements of LCA

7.2.2 Goal and Scope Definition

7.2.2.1 Goal

If you are to do an LCA, you should generally start by being as clear as you can about the goal, and writing it down. Recording the goal is both a requirement of the standard and necessary for efficient consulting practice, considering that clients may change the project manager assigned to your work or simply change their opinion about the goals of the work over time. Despite the fact that describing the goal seems simple and obvious, it is something which is frequently underemphasised in published LCAs, and something which leads to mismatches between the work done and the needs of stakeholders (Sandin et al., 2014).

The goal statement should say how the results of the LCA are going to be used in a broader context, why the study is needed and who the audience is. It is also important to note whether the intent is to use the results to make comparative assertions about products to the public. The latter places, according to the standard, the burden of independent peer review on the study. Of course, it may be worthwhile to get the work independently reviewed irrespective of this requirement.

One particular attribute should be made clear from the start: is the LCA to be *stand-alone* or *comparative*? Comparative LCAs (like the first one in Box 7.1) are often performed where the client wants to 'pick a winner' out of at least two possible options she could buy/construct/manufacture. Stand-alone LCAs (like the second one in Box 7.1) focus on one product or service and typically have the goal of finding out where the environmental 'hotspots' (key potential areas for improvement) are in the overall life cycle.

7.2.2.2 Scope
Functional Unit

The scope of an LCA report should tell the reader all of the key decisions that the analyst made in order to perform the calculations. The first of these is the definition of the functional unit. The functional unit is a description of the function or service

---

Box 7.1   **Examples of Simple Goal Statements in LCA**

1. 'The aim of the study was to help Sydney Water Corporation compare and prioritise a list of 10 possible infrastructure projects for the production of recycled water, in order to determine which should be funded. Stakeholders needing this information include the senior management team and representatives of the Department of Planning.'

2. 'This study was performed to estimate the carbon footprint of the Adidas cardboard packaging factory and enable the company to offset the footprint with an equivalent number of carbon abatement certificates. The immediate clients are the company's Finance and Environment Departments but since the results will be used in the annual report, an independent peer review is warranted.'

provided by the product or system under study. A functional unit should typically specify the quantity of the function (how much benefit is delivered), its lifetime (duration or durability) and its quality (reliability, usefulness, etc.). The point of defining a functional unit is to provide a basis for comparing two products or systems which may have very different components but are nevertheless able to provide the same function or service. As an idea, it is one of the most liberating aspects of LCA – since the focus is on a function rather than an object, the analyst can consider radically different ways of achieving the same ends. Some examples of functional units are provided in Box 7.2.

Students frequently have difficulty with the concept of the functional unit and confuse it with the units of impact indicators. Calling 'the kilograms of carbon dioxide equivalent emitted per plastic bottle' a 'functional unit' is an example of this mistake. Nobody wants the carbon dioxide to be emitted to the atmosphere – the emission is an unfortunate consequence of getting a function or benefit (in this case, the service provided by the bottle).

System Boundaries and Cut-Offs

LCAs are frequently focussed on products (e.g. plastic soft drink containers), production processes (e.g. sewage treatment) or services (e.g. delivery of telephone catalogue data). For the sake of simplicity, the rest of this text uses the words 'product' or 'product system' to describe all of these possibilities.

Once the goal and functional unit of an LCA have been described, it is good practice to draw a simple flow diagram for the product system to be studied, in order to explicitly consider where the system boundaries and cut-offs should be. Figure 7.3 is an example of such a preliminary flowchart for an LCA of a machine for turning atmospheric moisture into drinking water. A diagram like this helps the analyst think about the product system and what the key flows may be. Such a 'picture tells a thousand words' and is thus also an invaluable tool for discussions of an LCA study with clients and stakeholders who may have information about the scale of flows the analyst has identified, and knowledge of important flows the analyst has missed.

Box 7.2	Example Functional Units

1. 'Provision of 1,000 white cotton t-shirts'. This is a relatively simple functional unit which allows for different suppliers and production methods to be considered. Unfortunately, some shirts are so poorly made, they do not survive more than a few washes, so 'the use of a cotton t-shirt for every day for a year' is a better basis for comparing environmental outcomes. It would take into account that five low-quality shirts might be needed to supply the same function as one high-quality shirt.

2. 'The improvement of the city's water balance by an additional 10 ML/day'. This could allow alternatives such as rainwater tanks, seawater desalination and the reduction of leaks in the delivery system. This would not be the case if the same study used 'the delivery of an additional 10 ML/day into the city's water supply network'.

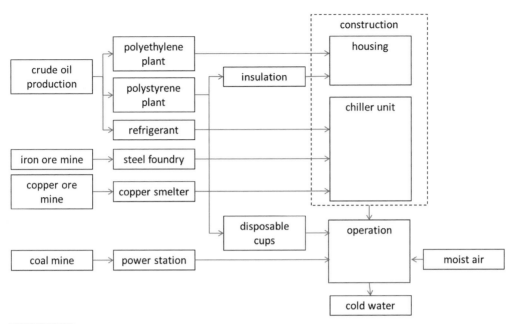

Preliminary flowchart for LCA of an 'air water generator' (Peters et al., 2013)

Whether something is missing is an important question for the analyst to ask, for example, 'how far up the supply chain for an ingredient in a food product is it appropriate for me to look?' In principle, an LCA should follow all inputs (resources, material and energy flows) back to the point at which they were extracted from the environment, and all emissions and products to the point at which they are discharged without further pollution control, but as with any engineering study, the person producing an LCA has to find an optimum degree of coverage and detail considering the intent of the client and the time available. At some point, once the most important flows have been considered, most supply chains are truncated or 'cut off' in LCA studies. In some standards, attempts have been made to quantify acceptable cut-offs numerically: PAS2050 sets a quantitative threshold of 95% of the total 'anticipated' emissions – these must be included along with 'any one source' that contributes at least 1% of the anticipated total. Cut-offs are acceptable in ISO14044 and should be made after consideration of their total mass or environmental significance. It is usual to exclude food consumed by workers and their travel to work from LCA studies, since one assumes, for example, that those people still have to eat even if the life cycle in the study does not occur. As a rule of thumb, it is more important to include raw material production for products which are 'passive' (e.g. furniture) than products which involve significant throughputs of materials or energy (e.g. water treatment plants, power stations), i.e. are 'active'.

A product system like the one shown in Figure 7.3, which condenses drinking water from the air, has various kinds of 'system boundaries', for example technical, temporal and geographic boundaries.

Technical boundaries include the boundaries between systems or within systems where multiple products are made. For example, in Figure 7.3 there will in practice be other co-products than polyethylene, polystyrene and refrigerant gas resulting from crude oil extraction (diesel fuel is an obvious example). So only some of the environmental impacts of crude oil extraction should be associated with the construction of the 'air water generator'. Dealing with this technical boundary is called 'allocation' in LCA, and is dealt with further in Section 7.2.3.3.

In an attributional (sometimes called 'retrospective' or 'accounting') LCA, the various parts of the life cycle that gave rise to the product are all considered. For example, a client wants to buy carbon credits to mitigate the damage to the climate caused by a product, and therefore has the goal of working out the scale of that environmental debt. On the other hand, a consequential (sometimes called 'prospective' or 'change-oriented') LCA is often performed to assess products that are not yet manufactured, or a change to an existing product system, where the question is 'what would happen if I did?' This may seem like a merely semantic distinction at first, but the implications for the life cycle inventory can be profound, and include whether the technical system is described using average or marginal data (see Section 7.2.3.5 for further discussion).

Another boundary of the technical system is where natural systems start. This is relatively easy to understand in the case of non-renewable resources like crude oil and iron ore. It requires more deliberation when the product system includes the extraction of water from rivers, or the deposition of waste in landfills. In the latter case, the analyst might be tempted to initially consider the landfill a permanent and sealed repository without connection to the environment, but will nevertheless have to consider the gaseous and liquid emissions. In the longer time perspective that some hazardous contaminants deserve, a landfill might be considered a part of the environment rather than part of a technical system.

Different temporal boundaries will be relevant depending on the goal of the study. For example, strategic decisions about future investments should naturally be supported by an LCA that is relevant for a long period into the future. On the other hand, if the LCA is comparing a mature technology with a technology that is undergoing rapid development, such a goal is in practice unobtainable, so the LCA report should try to describe the rate of change in the technology and focus on validity in the short term. If there are two alternative products in a comparative LCA, it makes sense to make the time span of the study greater or equal to the longer-lived product, so it is easy to see how many of the shorter-lived products are needed to provide the same service.

Geographic boundaries may also play an important role in an LCA, and therefore need description in the scope documentation. An obvious example would be the electricity supply shown in Figure 7.3 – someone in Sweden or France, countries where nuclear power and hydroelectricity provide low-carbon electricity, might react to this diagram because it indicates coal as the ultimate source of the electricity used during operation, but this particular study was performed in Australia, where the coal industry dominates electricity production. So describing the geographic boundaries of an LCA is an important way of delimiting the relevance of the results to different stakeholders.

Selection of Indicators

Although no results are shown in the goal and scope section of an LCA report, it is customary in this section to describe which life cycle impact assessment (LCIA)

Table 7.4 Strengths and Weaknesses of Alternative LCI Methods		
	Process analysis	EEIOA
Effort	LCI can be very time-consuming when many processes are involved, but the mathematics is relatively easy.	There is a large initial investment in setting up the EEIOA model, and the analyst has to be good at matrix algebra, but after such initial hurdles, analysis can be relatively rapid.
Accuracy	Nothing beats the accuracy of data collected from the process under study.	Using industry-sector average data reduces the potential accuracy of this method.
Coverage	Coverage of upstream processes can be difficult to achieve, resulting in early cut-offs and truncation error.	In principle, EEIOA avoids truncation error and is capable of following value chains of infinite length. Difficulties arise primarily when imports are a factor, though multi-regional EEIOA may overcome this.

supply (tonnes CO_2-e per kWh electricity) to estimate the contribution that part of the process makes to climate change, and added to the contributions made by other elements of the product system. This process analysis approach may be considered a 'bottom-up' approach, and the opposite of the 'top-down' input-output analytical approach. Input-output analysis (IOA) has its origins in the field of economics and is also performed outside the field of environmental assessment. Typically, the analyst builds a model of the entire economy of a nation and the financial transactions between industry sectors within that economy. When IOA is used in an environmental context ('environmentally extended IOA' or 'EEIOA'), environmentally relevant flows are aggregated for each industry sector so that the model contains data indicating a sector's total emissions. For example, in a study of the production of a t-shirt, one could estimate that a national knitting industry emitted a total of 24 kilotonnes of carbon dioxide in a year and produced \$16 billion worth of cloth (1.5 g CO_2/\$). The analyst then uses the IOA model to estimate (for example) that if \$20 was spent on the t-shirt, \$5 was spent in the knitting sector, and, using the corresponding average carbon dioxide emission for that part of the product system, this means that 7.5 g of CO_2 was emitted. Input-output analysis is discussed further in Section 7.7.

Process analysis and IOA have different strengths and weaknesses, as shown in Table 7.4. This has led to a number of attempts to hybridise these methods (see e.g. Alvarez-Gaitan et al., 2013).

7.2.3.3 Scaling and Allocation

Typically a product LCA will involve small parts of the total production capacity of certain systems. For example, a weaving facility will not use the entire electrical capacity of the regional power station. So the analyst will scale the total emissions of the power station by the proportion of its power output that the weaving factory uses to work out the emissions caused by the factory. More complicated problems arise

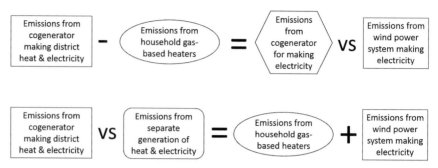

Figure 7.6   Examples of a system expansion

when processes in the product system have more than one product. Extending the weaving example, if the power station also supplies district heating as a co-product of generating electricity, then some of the emissions from the power station should be allocated to that co-product, reducing the amount associated with the electricity. Exactly how to do this has long been a topic of discussion and disagreement in the LCA community, but some consensus has arisen and has been incorporated in standards like ISO14044.

This ISO standard says, naturally, that the best option is to avoid allocation altogether. In some industrial processes, the problem can be avoided by subdivision, e.g. by improved monitoring at a larger number of locations in the industrial facility. Another way to avoid allocation is to 'expand the product system'. Figure 7.6 shows some examples of this. Imagine an analyst wishes to compare two possible product systems, one which obtains electricity from a wind turbine, and another which obtains it from a cogeneration facility which also delivers heat to houses via a district heating network. There could be a few ways to do this. Here are two. The analyst could add the function of delivering this heat to the wind power system by another means, say a simple gas burner in each house (system expansion by extending the scope of the functional unit – the bottom row in the figure). Thus it becomes possible to compare the cogenerator versus the alternative on a functionally equivalent basis. Alternatively, the analyst could subtract the impact of household gas heaters from the total impacts of the cogeneration process, to remove the difference between the functions of these two systems (system expansion by means of substitution – the top row). The analyst can then compare the result (the emissions from the cogeneration plant that are only associated with power production) with the emissions of wind power. (These are low, but a considerable amount of metal and concrete is involved in building wind turbines.) Of course, a great deal of care needs to be applied in identifying the correct process elements to use in system expansion – it has to be a realistic system expansion for the geographic and technical characteristics of the LCA, and the information and assumptions that led to the selection of the expanded system must be documented in the LCA report.

Where avoiding allocation is impossible, the alternatives are partitioning of the multifunction processes on a relevant physiochemical basis, or, if that is impossible, an economic basis. The simplest way to think about physiochemical allocation would be to partition an emission based on the relative flows of some quantity common to

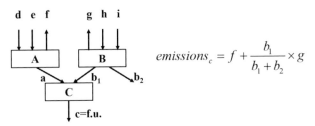

$$emissions_c = f + \frac{b_1}{b_1 + b_2} \times g$$

Figure 7.7   Allocation by physiochemical or economic partitioning

the co-products. Figure 7.7 illustrates this: if there are three processes A, B and C, and B produces two co-products b_1 and b_2, the analyst will have to allocate the emission flow 'g' between them in order to calculate the mass of emissions associated with the flow 'c'. It could be a matter of identifying the relative mass flowrates of 'b_1' and 'b_2' and using the formula shown. If the key benefit which the flows b_1 and b_2 provide is that they deliver energy, scaling the emission g by the total enthalpy of b_1 and b_2 makes more sense. (If neither mass nor energy nor other physiochemical parameters are known, the economic values of the two flows could be used.) Both of these possibilities are reasonable if producing 2 x c results in 2 x b_1, 2 x b_2 and 2 x g. On the other hand, it could also be that the causative relationship between c, b_1, b_2 and g is not proportional. In other words, if producing 2 x c means that 2 x b_1, 2 x g but still only 1 x b_2 are produced, then it can be argued that the emission g is not dependent on b_2, and the whole burden g should be allocated to c. The ISO standards ask the analyst not to slavishly allocate based on mass flows, but to base partitioning on 'the way in which the inputs and outputs are changed by quantitative changes in the products or functions delivered by the system'.

Economic allocation, the least preferred in ISO14044, may use a similar mathematical approach to that shown in Figure 7.7, but in this case b_i could be the financial value of the product streams b_1 and b_2, rather than a physical property of them.

7.2.3.4 Recycling Processes

Generally speaking, we can divide recycling systems into closed-loop and open-loop systems. In the former, the product can be remade back into the same product at its end of life. Aluminium beverage cans are an example of this. When a polyethylene terephthalate (PET) bottle is reprocessed into a polyester fleece garment, this is an example of an open-loop system. Allocation of impacts within an open-loop system is relatively complicated. Should the impacts of the extraction of crude oil be associated with the bottle or the fleece? Let us imagine a life cycle for PET with three use stages as shown in Table 7.5.

Several alternative approaches to allocation have been proposed (Baumann & Tillman, 2004; Ekvall & Tillman, 1998), and some of the most popular ones are summarised in Figure 7.8. The simplest allocation procedure is '*cut-off*' allocation – each product is associated with only the emissions that were required to produce it, except for the last product, which also caused the emissions associated with waste disposal. This may make sense if it seems there is no overall governance where the

Table 7.5 Imaginary PET Life Cycle Data

Life cycle stage	Quality (index)	Emissions (kg CO$_2$/g)
Original manufacturing (e.g. crude oil extraction, bottle blowing)		5
Use of PET bottle (the first product: 'P1')	1	
First recycling (including spinning) processes		2
Use of PET fleece (the second product: 'P2')	0.75	
Second recycling (including moulding) processes		4
Use of plastic noise barrier (the third product: 'P3')	0.5	
Waste disposal		1

system is located, and each product user can only think of themselves. This is similar to *extraction loading* except, in this case, the argument is that the ultimate waste is a consequence of the original resource extractive activity, so the impacts of waste disposal are associated with the first product. Since this means subsequent users of the recycled material are not burdened with its initial extraction, this encourages additional recyclers to join the system. Another view is that whoever is ultimately responsible for waste disposal is responsible for the impacts of resource extraction –

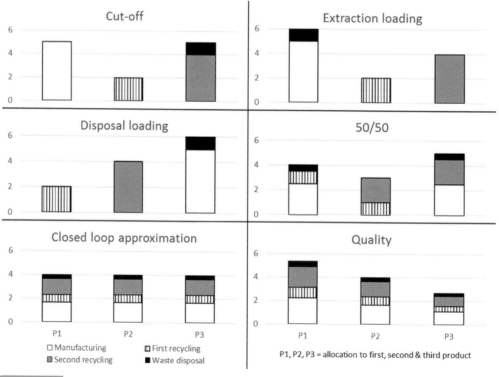

Figure 7.8   Alternative approaches to allocation for open-loop recycling (scale = kg CO$_2$)

so *disposal loading* provides the current final user with an incentive to ensure their product can be recycled. A compromise between these views is to say that both production and waste disposal are necessary for each product, and to associate half the impacts of a process with the previous and subsequent products in the system. This has been called *50/50* allocation. If we take that perspective to the extreme, we may say that all the products are responsible for a fraction of all the impacts of all the processes, warranting allocation by *closed-loop approximation*. Figure 7.8 shows that, in this case, each product has the same impact – a third of the total. A more subtle version of this perspective is to say that the overall system is driven by the value that each product provides. In this case, we might allocate the total environmental impacts in proportion to the quality of each of the products. For PET, this might be the material's purity or tensile strength, or if no better proxy can be found, the price per kilogram. (Quality is represented by a dimensionless index in Table 7.5.)

7.2.3.5 Attributional and Consequential Perspectives

Along with allocation, the other area of LCI methodology that commonly challenges analysts is the question of whether to take an attributional or consequential perspective. The first LCA studies were typically attributional – they answered the question of what environmental impacts were caused by industrial systems making a certain product. For this reason they have also been called 'accounting type' or 'retrospective'. On the other hand, sometimes we are more interested in questions like 'what if ... ?' For example: how would the environmental impact of a system change if we were to increase the proportion of recycled materials going into it? Such studies are focussed on the consequences of a decision and have therefore also been called 'change oriented' or 'prospective', although they do not necessarily have to be based on a future time.

The difference between attributional and consequential LCA can be illustrated at various levels of complexity. The simplest is shown in Figure 7.9. Imagine the analyst is performing an LCA to understand the total impact of her breakfast on climate change. If she eats weetbix, the relevant system is shown as 'option A' in the figure. All six small boxes are relevant, and it is likely that adding milk to the weetbix will

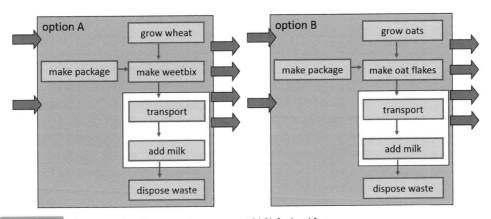

Figure 7.9   Eliminating shared processes in consequential LCA for breakfast

cause an important part of the impact, due to the methane that cows generate. This would be an attributional LCA, looking back at where her breakfast came from. On the other hand, if the analyst wants to consider changing her diet to option B (oat flakes), this is a different question and the assessment of the weetbix system may be different. In particular, the milk may be irrelevant if the same volume of milk is needed to make the breakfast edible under both options. Likewise, if the transportation has about the same impact, since the cereal has about the same mass and density, this might reasonably be left out of a consequential analysis. Now we only need the parts of the system on the grey background and can leave out the two steps in the white boxes.

At this point it may sound like consequentialism makes the analyst's life easier and the impact statistics all lower, but consequentialism often does the opposite. Ekvall et al. (2005) present a nice example that seems as simple as the one given earlier, but raises further questions. Imagine a hotel on an island that has its own micro-hydroelectric power supply. It also has a cable to the mainland, to allow it to sell the low-carbon electricity when it has an oversupply. A university professor is wondering whether to organise a conference at the hotel. From an attributional point of view, this would be a good thing – the impacts of the electricity supply will be close to zero. From a consequential point of view, however, sending the conference delegates to that hotel will increase electricity consumption there, reduce the sale of electricity by the hotel into the grid, and probably mean that somewhere else, more gas or coal has to be burnt to make up the difference. So this is an example in which a consequential perspective indicates a higher impact.

The hotel example also demands more data. A key challenge is to identify what the marginal technology will be. Just how far into the wider world does the analyst need to look to find it? (In fact, the same problem is present in the breakfast example: what happens to the cereal the analyst does not choose? However, it is unlikely that one individual's diet will influence the planning of industrial agricultural systems managed by other people.) Wiedema et al. (1999) elaborate the following stepped process to help to identify the systems affected by a choice in a consequential LCA:

1. Determine what time horizon is relevant. If it is short term, the LCI may only need to use existing industrial capacity. Otherwise, construction of new infrastructure (new capacity) may be warranted.
2. If the time horizon is long term, identify whether the change affects specific processes connected with the system under study (in which case they are the marginal processes), or the broader market for the goods it provides.
3. If the broad market is affected, ask whether the supply is increasing or decreasing. If it is decreasing, the marginal technology is probably old and inefficient. If it is increasing, the marginal technology is probably newer and more competitive.
4. Work out whether the marginal technology offers the potential to provide the desired change in the volume of production. (E.g. there may be absolute, political or technical limits to supply. The potential supply of manure is not much affected by the demand for fertiliser – manure is a by-product of the demand for food.)
5. For whichever technologies are not limited, identify which one the decision-maker will prefer to build (typically the cheapest or safest, depending on the type of technology) or shut down (probably the most expensive or risky).

Can it get more complicated than this? Yes it can! Consequential LCA can be described as an attempt to integrate economic modelling into LCA. In a number of studies, analysts have used more advanced economic modelling techniques, like partial equilibrium modelling and computable general equilibrium modelling, to create the LCI needed for an LCA and to thereby give it consequential characteristics. This book does not attempt to explain these approaches, but for the interested reader, the review by Earles and Halog (2011) is a good starting point.

7.2.3.6 A Note about Scopes in the GHG Protocol

If you spend any time in your career working on LCAs or public environmental reporting (see Chapter 8), you are bound to hear people talking about to what extent you included 'scope three'. This term has become commonplace due to the development of the Greenhouse Gas Protocol by the World Resources Institute and the World Business Council for Sustainable Development. The Protocol is in fact three separate standards for corporate accounting, project accounting and product accounting, and to help to keep them consistent and easy for readers to understand, the authors found it useful to describe three separate 'scopes' of emission sources when describing what analysts should calculate.

'Scope one' refers to the emissions coming directly from a business subject to a greenhouse gas calculation, that is, the gases physically originate from the business' premises. 'Scope two' refers to the emissions caused by the energy purchases (electricity, steam, etc.). The actual emissions come from another company (e.g. the local power company), but it is relatively easy for analysts to get hold of data enabling these emissions to be calculated. 'Scope three', on the other hand, is everything else: emissions caused by the use of electricity by suppliers, emissions physically originating from the premises of suppliers of materials and services and the like. Figure 7.10 illustrates

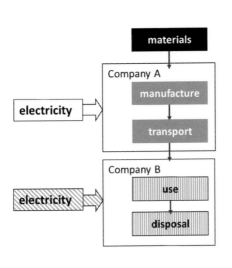

Figure 7.10   Which 'scope' is my emission in? Comparison of two companies

how this would apply to two fictional companies A and B. Notice how the emissions that are in scopes one and two for company A are in scope three for company B.

There is a great deal of debate in corporate greenhouse gas accounting (where the object under study is a company, rather than a product) about how much of the scope three emissions must be included, if any, when a company reports its emissions. On the other hand, when the object under study is a product or service, LCA analysts have tended to be more concerned about whether a particular source of emissions is relatively large or small in the context of the total sum of emissions, rather than who owns the source (see cut-offs in Section 7.2.2.2).

7.2.4 Life Cycle Impact Assessment

7.2.4.1 Classification and Characterisation

According to the ISO standard, classification and characterisation are the compulsory parts of LCIA, the third stage in an LCA. Classification means associating individual items from the LCI with the kinds of impacts they cause. These may be various impacts on human health, different kinds of damage to environmental quality or one of a number of types of resources (e.g. land, water, etc.). In many cases, a particular emission may cause more than one kind of impact. For example, laughing gas (nitrous oxide, N_2O) is an emission that can potentially cause eutrophication and stratospheric ozone depletion, and it is also a powerful greenhouse gas.

If several emissions are classified as causing the same kind of impacts, the challenge arises of how to aggregate them to simplify the information in a meaningful way for decision-makers. By characterising emissions, the LCA analyst converts the emissions to a common scale. The most well known of these scales is the idea of 'carbon dioxide equivalents' (CO_2-e) for greenhouse gas emissions. These are based on the extent to which different gases impact 'radiative forcing' – the difference between the amount of energy the Earth receives from the sun and what the Earth radiates back to space.

Table 7.6 Characterisation Factors for Global Warming Potential (GWP) (Adapted from IPCC, 2014b)			
Greenhouse gas	Atmospheric lifetime (years)	GWP_{20} (kg CO_2-e/kg)	GWP_{100} (kg CO_2-e/kg)
CO_2	*	1	1
CH_4	12.4	84	28
N_2O	121	264	265
CF_4	50,000	4,880	6,630
HFC-152a	1.5	506	138

* Since carbon cycling in nature is the result of many different dynamics, the IPCC does not propose a single figure here. On average, CO_2 molecules are exchanged between the atmosphere and the earth every few years, but it will take a few hundred thousand years for the carbon to be reabsorbed into the lithosphere.

Table 7.6 shows some data from the latest (fifth) assessment report from the Intergovernmental Panel on Climate Change (IPCC, 2014).

In an LCIA, these values ('characterisation factors') can be used to characterise greenhouse gas emissions, so if the life cycle of a t-shirt causes the emission of 1 kg each of CO_2, CH_4 and N_2O, the contribution to climate change of the t-shirt is 294 kg CO_2-equivalents (using a 100-year timeframe). One number is easier to communicate and compare with others than three numbers. This aggregated value can be called a 'greenhouse gas LCA' or a 'carbon footprint'. Note how in the table there are different values for the 100-year timeframe and the 20-year timeframe (GWP_{100} and GWP_{20}, respectively). These timeframes represent the relative importance of these gases and the fact that it depends on how far you look into the future. If you are most concerned about the short-term effects on planetary life of greenhouse gas emissions, then a gas which per kilogram has a more powerful insulating effect on the atmosphere than CO_2, for example CH_4, is more important. On the other hand, the impact of CH_4 decreases rapidly as it reacts with hydroxyl radicals in the atmosphere to produce CO_2, resulting in an atmospheric lifetime of about 12 years. CF_4 has, on the other hand, an extremely long atmospheric lifetime, so the characterisation factor for a 100-year perspective is higher than the equivalent value for a 20-year perspective. Note that these data have been updated between IPCC reports, for example scientists have steadily decreased their estimates of the impact of nitrous oxide emissions, and increased them for methane. The Kyoto Protocol used 21 kg CO_2-e/kg as the GWP_{100} value for methane, based on the IPCC's second assessment report from 1995, and at the time of the writing of this book, many people were still quoting the fourth assessment report (2007 data) with its 25 kg CO_2-e/kg CH_4.

7.2.4.2 Midpoint and Endpoint Indicators

An aggregated indicator for contributions to an environmental impact based on a common reference chemical (such as a climate change indicator based on GWP_{100}) is a practical way to describe a large amount of data with a single figure. Such indicators are often called 'midpoint indicators', as distinct from 'endpoint indicators'. Figure 7.11 provides an introduction to the difference between them. If an analyst wants to compare alternative vehicles for getting from A to B using LCA, she will probably find there are carbon dioxide and methane emissions somewhere in the life cycle of a standard car. By identifying how far the car will travel using a map, and referring to laboratory tailpipe data on vehicle emissions, she should be able to begin an LCI. The international consensus around characterisation factors described previously will enable her to aggregate these to a single figure. Having thus characterised the emissions, she can be said to have performed an LCA, but the results only describe the impact in terms of their equivalent contribution to radiative forcing. This may be too abstract for decision-makers. Most people do not have any idea about how important 1 tonne of CO_2 is, or whether 1 W/m^2 is a lot of radiative forcing. For them, the more interesting questions may be: 'How much warmer will the planet get?' 'Will my parents die early due to heat stress?' and 'Will I get to see the Great Barrier Reef before it is bleached to extinction?' Such matters lie further along the cause-effect chain from emissions, mediated by variables of atmospheric physics, and in the case of coral bleaching, oceanography and ecological responses to heat stress.

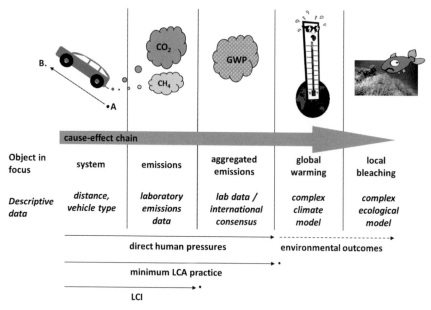

Figure 7.11 Midpoints and endpoints in LCIA

Figure 7.12 shows an example of a popular, integrated LCIA method in which most potential impacts have been described using characterisation factors at both the midpoint and endpoint levels. The 'Recipe' method (Goedkoop et al., 2008) enables analysts to aggregate 16 different kinds of impacts into endpoint indicators for three overarching 'areas of protection': human health, ecosystems and resources. So for example, the analyst can describe the climate impacts of the production of a t-shirt, either in terms of a midpoint indicator for radiative forcing (in kg CO_2-e), or in terms of the endpoints of (a) the elimination of species integrated over the time they are eliminated, and (b) the human Disability Affected Life Years (DALYs) (a value defined in Chapter 6) associated with the change in climatic conditions. The units of the other midpoint indicators are shown in the centre of the figure (DCB = dichlorobenzene; PM10 = particulate matter between 2.5 and 10 µm; NMVOC = non-methane volatile organic carbon; CFC-11 = chlorofluorocarbon 11; U235 = uranium 235).

The use of midpoint indicators has the benefit that the data are much more certain. Building a model of the likelihood of coral bleaching involves many more dynamics and assumptions about adaptation than presenting a midpoint indicator. On the other hand, for some audiences, a more uncertain but palpable indicator such as DALYs may be more meaningful, and for many audiences, the fewer the indicators the better.

LCA researchers have borrowed heavily from other scientists, not just the climate experts of the IPCC, to come up with the various models in use to calculate life cycle impacts. The USEtox model introduced in Chapter 4 is an example of a multi-box model of the environment which was originally designed for chemical risk assessment, but has been adapted over many years to calculate characterisation factors for the assessment of toxicity in an LCA framework.

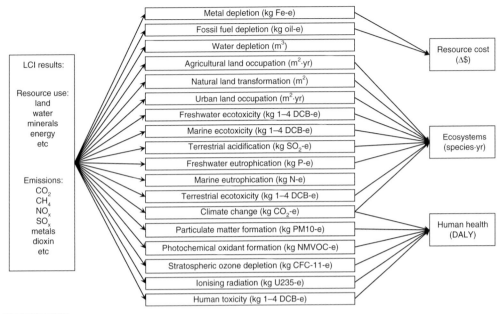

Figure 7.12 Midpoint and endpoint indicators in the Recipe LCIA method (Goedkoop et al., 2008)

7.2.4.3 Normalisation and Weighting

While using an endpoint indicator system like Recipe is intended to reduce the number of indicators a decision-maker has to consider, some stakeholders only want a single figure for 'the environment' in a decision that may involve many other parameters. Under these circumstances, the LCA analyst may wish to engage with two elements of LCA that are optional according to the ISO standards: normalisation and weighting. These may be applied to endpoint indicators, or to midpoint indicators, depending on the decision at hand and the LCIA method in use.

Normalisation refers to the comparison between an indicator and some kind of external reference. For example, some LCIA methods normalise the emissions of a product life cycle against current global emissions of the same kind. This may help to prioritise issues for policy intervention. Normalisation is of itself less open to criticism of subjectivity than weighting, since the normalisation reference points are typically measurable, physical values. On the other hand, there may be debate about whether local, regional, national or global reference points are more relevant in a particular case.

The more fundamental criticism of normalisation is that by itself it does not indicate the importance of the emissions associated with a product. If a product causes 1% of the emissions that cause two global environmental problems, it is still unclear whether one of the problems is more important. Weighting methods have been developed to help to deal with this issue. The process of weighting LCIA indicators can be based on many different information sources. Notable methods have used citizens' willingness to pay (EPS) (see Steen, 2016), political environmental goals (EDIP) (see Wenzel & Alting, 1999) and the values of expert panels (EI99) (see Goedkoop & Spriensma, 2000). Essentially, weighting is a multicriteria analytical (MCA) process. MCA is described in more detail in Chapter 9.

7.2.5 Interpretation

The fourth stage of the iterative LCA process is called 'interpretation', and it is an activity that is relevant to all three of the other stages. The primary focus of the interpretation stage is usually a *contribution analysis*, which is when the analyst examines which flows noted in the LCI were the key determinants of the outcomes of the LCA. Typically, a few of the emissions or resource flows will cause most of the potential impacts identified in the LCIA. They may arise from only a small subset of the processes that make up the product life cycle. Knowing which flows and which processes are the most important enables the LCA analyst to provide advice to decision-makers on where they should focus their efforts towards improving product environmental performance. For this reason, this stage of the process was once called 'improvement analysis'. Of course, identifying the major contributors can sometimes reveal errors, where flows were larger than expected for no good reason, so it can also help to improve the quality of the analysis.

7.2.5.1 Completeness and Consistency Checks

Another key element of the interpretation of an LCA is quality control. Once the analyst has something written down in a document for all the previous three stages, and can thus place them side by side, it can become apparent that there are inconsistencies between the parts which need to be rectified. In a *completeness check*, the analyst will examine whether some parts of the system suggested in the goal and scope have somehow been forgotten. She may also realise that while some kinds of information about the study may still be in the analyst's mind, they have not been written down, preventing a thorough interpretation of the results by anyone except the analyst. A *consistency check* will identify whether the assumptions, methods and data are aligned with the goal and scope of the LCA, for example, whether the generality of the data describing a process was actually appropriate for the intended goal (see Section 7.2.2.2).

7.2.5.2 Sensitivity Versus Uncertainty Analysis

Sensitivity analysis and uncertainty analysis are both worthwhile elements of quantitative analysis, and are often employed in LCA. They are similar ideas in practice, intended to identify the influence of variations in process data, modelling choices and other data. Arguably, the difference between them is that in sensitivity analysis (sometimes called 'what-if analysis'), the analyst deliberately introduces variations in the input variables to determine what the main sources of variability in the outcomes are. So, for example, one might look at Box 7.4 and vary the inventory data and the normalisation data up and down by a standard value of 25%, to see what effect each variable has on the bottom line. Uncertainty analysis does the same thing, but typically uses empirical data on the uncertainty ranges of specific variables in order to calculate a total error range in the results. For example,

Box 7.4 **Example of LCIA for Some Functional Unit (FU) Emitting N$_2$0 and CO$_2$**

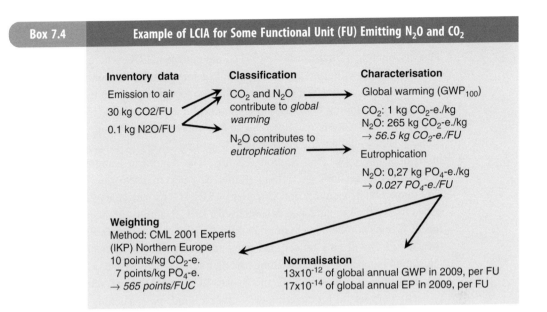

one might know that the inventory data are within 10% of the likely true value, and that the normalisation data are within 50%, so these uncertainty margins are used in the calculations.

7.3 Other Key ESA Tools

LCA is not the only possible assessment tool for analysts who want to consider environmental impacts, and it is related in various ways to some tools concerned with issues other than the natural environment. Some of these relations sound obvious from their names: life cycle costing (LCC) and social life cycle assessment (SLCA). Others are connected instead by sharing with LCA one or several similar mathematical steps, for example Material Flow Analysis (MFA), Material Intensity Per Service (MIPS) and input-output analysis (IOA). Here we build on the understanding of LCA you have by now, to briefly explain what each of these other analytical tools is.

7.3.1 Life Cycle Costing

Life cycle costing is more closely related to the process of cost-benefit analysis (CBA) than it is to LCA because the focus of LCC is on calculating a financial net present value (NPV) of alternatives, rather than calculating environmental indicators. The basic idea of NPV is shown in the following equation.

$$Net\ Present\ Value = \sum_{t=1}^{T}\left(\frac{(B-C)}{(1-x)^{t-1}}\right)$$

In this equation, B is a series of benefits and C is a series of costs coming at some time-step (t) between the start and the maximum lifetime T, while x is a discount rate. This discounting of future cash flows is commonplace in LCC, whereas LCA experts are generally unwilling to apply a discount factor to future environmental impacts. The principal exception is that in some LCA models of impacts on human health, an age-based discounting of effects on human health can be applied (Kobayashi et al., 2015). LCC calculations attempt to be more comprehensive than some NPV calculations by taking a longer life cycle perspective of products and services and including more of scope three (see Section 7.2.3.6). The American National Institute for Standards and Technology (NIST) (Fuller & Petersen, 1996) and the Society of Environmental Toxicology and Chemistry (SETAC) (Swarr et al., 2011) have published two of the key guidance documents for LCC. Naturally, if the person performing LCC does not include environmental externalities among the values of B and C, then the LCC will only reflect environmental costs which have been monetised by legislation (such as laws establishing carbon trading schemes) and/or legal processes (such as fines for oil spills). In this sense, LCA and LCC are very different and typically very useful for covering each other's blind spots. There is a risk that if an analyst only relies on LCC, only a few, if any, of the environmental consequences of a decision will be considered.

7.3.2 Social Life Cycle Assessment

SLCA is a new field compared to LCA and LCC, and therefore it is the subject of much methodological dynamism and uncertainty. It shares the same four conceptual steps as LCA but introduces different damage indicators. The UN Environment Programme (UNEP)/SETAC proposed a list of SLCA indicators based on the kinds of stakeholders affected by product systems (workers, value-chain actors, consumers, local communities and general society) and the kinds of social impacts that seem relevant (working conditions, child labour, poor salaries, corruption, etc.) (Benoît & Mazijn, 2009). Since the list is long, analysts still have to identify which indicators to use, and this is the subject of much critical discussion (Arvidsson et al., 2015; Sandin et al., 2011). One of the key issues in SLCA is that identifying that a product comes from a place where social problems exist (e.g. poor employment conditions in textile workshops in a south-east Asian country) does not necessarily mean that when a European or North American buys a shirt from such places, they contribute to the problem. Sometimes the opposite may be true. The possibility of predicting or estimating social impacts in product life cycles is keenly sought by many people.

Since the concept of sustainability traditionally covers impacts not just on nature, but also on human society and the economy, SLCA and LCC are sometimes grouped together with environmental LCA under the heading of life cycle sustainability assessment (LCSA) (Kloepffer, 2009).

7.3.3 Material Flow Analysis

In material flow analysis (MFA) (also called *material flux analysis*), an account is typically generated for the whole or parts of the economy of a region or a country.

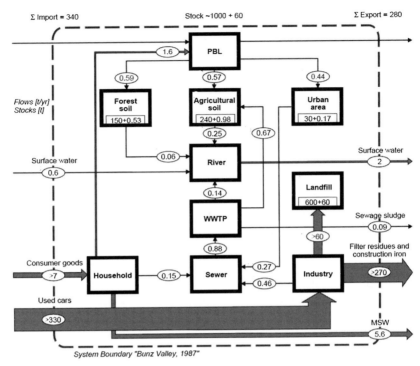

System Boundary "Bunz Valley, 1987"

Figure 7.13   MFA on major lead flows in a region of Switzerland (Brunner & Rechberger, 2005, reproduced with permission)

The first step is to define how a region or system is to be studied. Figure 7.13 shows an example of MFA applied to major lead flows in a region of Switzerland. The scale of mass flows in and out of the system, the accumulation of stock and possibly also flows between units within the system are estimated. The level of detail varies greatly from MFA to MFA and the types of flows that are considered may be anything from total mass flows (in e.g. material flow accounting for a nation) or single substances (such as in substance flow analysis [SFA]). With its emphasis on mapping flows in an economy, MFA is typically used to inform regional and international environmental policy development rather than the design of industrial products. MFA is similar to LCI in that mass (and possibly also energy) exchanges are mapped out. Therefore MFA data are relevant to LCA analysts. They may want to focus on an industrial production chain rather than a geographical region, but they can nevertheless learn about where material flows go from MFA analysts.

7.3.4 Material Input Per Service and Ecological Footprint

Material Input (or 'Intensity') Per Service (MIPS) is a calculation that may be useful to product designers who want to do an LCA but are happy to settle for the total mass of material associated with a product life cycle, as a sort of proxy indicator for many environmental damages. MIPS is essentially an unweighted single-indicator LCI result based on an MFA. MIPS has been developed and promoted by the Wuppertal Institute in Germany (Ritthoff et al., 2002). The Institute provided data for the calculation of five categories of material which are used in the production of different inputs to a product or service: abiotic raw materials (minerals, energy carriers, excavated soil), biotic raw materials (wild and cultivated biomass), water, air and agricultural soil (moved in some way; includes tilled and eroded soil). Ritthoff et al. (2002) give an example in which they add all the solid materials (abiotic, biotic and soil) to arrive at an 'ecological rucksack'. Table 7.7 shows some examples of the kinds of factors recommended (WICEE, 2014).

A practical strength of MIPS is the speed with which an analyst can calculate it compared to a full LCA. The results can also be easy to communicate. On the other hand, the method assumes that the impacts of products and processes are proportional to the total mass of material they disturb, which is obviously an approximation.

A further example of an extension of MFA is when different types of flows are translated into the area of land used to provide the flows, which is a step in calculating an ecological footprint. Ecological footprints have been promoted by the Global Footprint Network, the brainchild of Mattias Wackernagel, who along with William Rees developed the idea of calculating a single figure for human demands on ecosystems, in hectares. Like ecological rucksacks, ecological footprints have the advantage of being easier to communicate than other environmental performance metrics. It is hard for the average person to know if a tonne of carbon dioxide is a lot or a little, but you can compare ecological 'global hectares' with the number of hectares on planet Earth, or the number of hectares of land around a typical house in your city, to get a feeling for the scale of an ecological footprint. The downside of course, is that a lot of information gets lost in turning a complex picture into a single

Table 7.7 Examples of Material Intensity Factors (kg/kg Except Where Noted*)					
	Abiotic material	Biotic material	Water	Air	Soil
Cotton (United States)	8.6	2.9	6814	2.74	5.01
Container glass (Germany)	3.04	-	17.1	0.72	0.14
Spruce timber (cut, dried, Germany)	0.68	4.72	9.4	0.16	-
Paper (bleached, EU)	9.17	2.56	303	1.28	-
Polypropylene (injection moulded, EU)	4.24	-	205	3.37	-
Truck transport* (kg per tonne-kilometre, Germany)	0.22	-	1.91	0.21	-
Electric power* (kg per kWh, EU average)	1.58	-	63.8	0.43	-

figure, so these data are generally used for policy evaluation at the regional and national levels, rather than for comparing alternative designs for consumer or industrial products.

7.3.5 Input-Output Analysis

IOA was first (and is still primarily) applied to assess economic impacts. It was originally developed by Wassily Leontief in the 1930s (who received a Nobel Prize for this in 1973). There are a number of very practical descriptions of how to do IOA (Hawkins & Matthews, 2009; Kitzes, 2013). Typically, annual flows between industry sectors in a national economy are gathered in an n x n data table, where n is the number of industry sectors in the economy. The data in this table are normally denominated in terms of the currency of the particular national economy. The table is thus effectively a high-level map of all the financial transactions in an economy. Data on such flows are today widely available from national statistical agencies. By inverting the matrix, the analyst can calculate how a dollar of expenditure in one industry (e.g. paper making) ultimately ends up spread across the economy (e.g. into sectors like forestry, mining, electricity generation, etc.).

IOA became more useful for environmental assessment when in 1970 Leontief suggested extending it to assess the amount of pollution associated with production and/or consumption. To do this, the monetary input-output table needs to be complemented by an n x 1 vector providing the mass of pollution associated with a currency unit (e.g. dollar) of production by each industry sector (e.g. kg CO_2 per dollar of paper products). By converting the output of an IOA in this way, an environmentally extended IOA (or EEIOA) is obtained, which can present for example the total greenhouse gas emissions associated with a dollar spent on paper making. Similar efforts have been made in SLCA to associate social impacts (and benefits) with a dollar spent in different industry sectors.

From an LCA analyst's perspective, EEIOA is a neat way of avoiding complicated and time-consuming efforts to track material flows during the LCI. Relatedly, this can help the analyst avoid 'truncation error' – missing flows and impacts in background parts of the system that are laborious to enumerate because they are distant from the actor commissioning the LCA study. On the other hand, EEIOA is built on industrial sector averages calculated by statistical agencies – if the segmentation into industry sectors is too coarse (i.e. not many sectors are defined), this can lead to large errors. For example, if all mining activities are aggregated in a single industry sector, an EEIOA of a product with low-impact mined inputs may be biased by another particularly environmentally damaging mineral in the same sector – in an economy with only gold and coal production, the carbon footprint of a dollar spent on gold will appear to be the average of the footprint of the two mined products. Many analysts try to get the best of both worlds by hybridising IOA and traditional process-analysis-based LCA – using process data to estimate the impacts of the foreground system where engineering data are relatively abundant and of higher quality than IOA data, and using IOA for distant parts of scope three of the system where the relationships between actors are otherwise unknown.

7.4 Strengths and Weaknesses of ESA Tools

The development of IOA, MFA, MIPS, LCA, LCC, SLCA and LCSA represents attempts to expand the thinking of decision-makers from a narrow focus on money (e.g. NPV) when developing products and policies, to give decision-makers the possibility of a better informed and broader worldview. These tools include a larger vision of the technical system that delivers the products and services we ask for, and a broader conception of the environment in which those products and services are created. As a consequence, they are better at avoiding problem-shifting than assessment tools with a more limited perspective. They can also help us identify trade-offs between different dimensions of sustainability. Beaulieu et al. (2015) suggested that these tools and others can help direct the transition of our economy towards sustainability. Finnveden and Moberg (2005) provide a useful review of ESA tools. Beaulieu et al. (2015) also provide a lengthy discussion of the relationship between these and other tools.

Of course, none of these tools is the 'silver bullet' that will provide all the information we need to guide decision-makers. A common critique against all systems analysis is that it presumes that a system can be mapped and analysed in all important respects. Systems analysis is sometimes claimed to be an expression of a reductionist view of the world in the sense that it attempts to explain systems in terms of their individual, constituent parts and interactions between them. Further, it often also relies on a positivist paradigm, prevailing in natural sciences, in which knowledge is based on what can be observed and measured. It is important to understand the limitations of any systems analysis tool that is used in a specific context.

Andersson et al. (2014) describe different types of (societal) systems as simple, complex, complicated or both complex *and* complicated, the latter referred to as *wicked*. They suggest that simple systems can be described by mathematical theory whereas complicated ones (such as computers or technical systems with many different components) need tools developed within the field of systems-based theories. However, when it comes to complex systems (with components that are strongly interconnected, such as the flight path of a flock of starlings), these are dealt with by completely different tools developed within complexity science. But for the wicked systems that exhibit characteristics of both complicated and complex systems, neither of these types of tools is sufficient and different approaches are needed.

It is sometimes argued today that many of the problems that engineers will encounter in their professional life actually concern wicked systems, but that engineers are rarely trained to manage problem-solving in such situations. In fact, they may not even be aware that systems may be wicked and may therefore expect that more traditional tools can be used in such situations. Exactly what approaches should be used to handle wicked problems is a matter of almost tautological debate – the problems would not be wicked if there was an easy solution. Nevertheless, one characteristic of approaches for handling wicked systems is that they should assume emergence (that the system changes and achieves new properties upon manipulation), that they iteratively address different levels of the system (micro and macro levels) and operate iteratively as new information arises and the system is transformed, that they accept that information and the ability to represent system dynamics is incomplete, and that stakeholders have differing views and can change their minds about what matters to them. Arguably, this does not rule out the use of ESA tools, but demands that they be used to inform an ongoing dialogue rather than imagined as instruments of control.

In conclusion, in the context of sustainable development, the need for more holistic and broader considerations to avoid e.g. problem-shifting and suboptimisation pushes us to increasingly use systems analysis and even apply more advanced systems analysis methods that consider multiple aspects of complex systems. However, we always need to be aware of the limitations of such analytical tools and be prepared to use complementary approaches to understanding and managing real-world problems. For wicked systems, LCA and other system analytical tools may be best thought of as instruments of learning and dialogue with stakeholders, rather than providers of absolute results.

7.5 Review Questions

1. What issues and questions encouraged the development of life cycle assessment in the 20th century?
2. Think of a key environmental problem in your home town. In what decision situations would a life cycle perspective be necessary to solve it?
3. Consider a product in your home and an alternative to it that can deliver the same function. Now imagine you are working for the company that manufactures this product, and looking for a marketing edge over the alternative. Draw a preliminary flowchart for the product system. Write a goal and scope which you can use to subcontract the LCA work.
4. Write down an alternative functional unit to the one you created in question 3. What are the strengths and weaknesses of the new one compared with your original?
5. Which impact categories would be relevant to the product in question 3?
6. How can different temporal frames of reference affect how an LCA is done?
7. What would be the advantages and disadvantages of doing EEIOA rather than a standard process LCA of the product in question 3?

8

Regulatory Structures

In this chapter, we look at how environmental management works in an organisational sense – what instruments are used to manage the environment in public and private organisations at various levels of geographic coverage. The chapter begins with international environmental legislation, examining key examples and their focuses. Below this level is a whole world of governments for nations, parts of nations and groups of nations, so the chapter focuses on just a few that have international ramifications for engineers. These are: the European Union (EU) – an example of a multinational jurisdiction, and a massive and wealthy market which led the world by introducing the REACH legislation on chemicals; the United States of America – an example of a single nation that is also one of the biggest markets in the world and one of the pioneering jurisdictions; and Sweden – which, perhaps by virtue of its small size and tradition of large government engagement in the economy (see Figure 8.1), has led the EU on some fronts of environmental management. Finally, we look at the corporate scale – the company as a self-governing entity – and three common key elements of environmental governance in which corporate entities engage: environmental impact assessment, public environmental reporting and environmental management systems.

8.1 International Environmental Laws

Many chemicals are very persistent. Many persist long enough to be transported by winds, ocean currents or even migrating animals to countries and continents far from their sources. Therefore, a country cannot protect itself from some environmental problems without the help of its neighbours and friends around the world. This observation has motivated a number of international environmental laws. Some of the key examples described here relate to persistent organic pollutants (POPs) (the Stockholm Convention), long-range air pollutants (CLRTAP), ozone-depleting substances (the Montreal Protocol), greenhouse gases (the Kyoto Protocol and the Paris Agreement) and hazardous waste (the Basel Convention).

The making of international laws can be a slow process, and generally has three stages. The first is the negotiation process, in which a country sends a delegation of representatives to meet delegations from other interested countries. If they have a successful meeting, this process ends with the approval and signing of a common document by the leaders of the delegations. Signing indicates the intention of a signatory country to comply with the treaty, but it usually means a second stage is required: ratification. This is where the delegation comes back to its country of origin

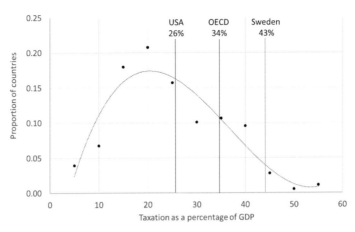

Figure 8.1 Tax as a proportion of gross domestic product is one measure of the scale of government engagement in managing a national economy. The OECD is a forum whose members include 35 of the wealthiest democracies (2017 data: from www.heritage.com)

and tries to get the national legislature to approve the treaty. For example, in the United States, a two-thirds majority of the Senate is required to ratify a treaty even if the president has already signed it. This may mean passing a new national law reflecting the central provisions of the treaty. The third stage is implementation, when the country tries to make the law work in practice. This may mean enacting detailed regulations or additional laws and pursuing other government initiatives.

8.1.1 Stockholm Convention

The Stockholm Convention is a treaty negotiated under the auspices of the United Nations (UN). It was signed in 2001 and came into force in 2004 when 50 nations had ratified it. It has been widely adopted, with 180 signatory nations in 2017 (compared with a total of 193 members of the UN). Notable exclusions include the United States, Italy and Malaysia.

The Convention is focussed on POPs – the initial 'dirty dozen' persistent organic pollutants (see section 3.7) banned by the Convention are mostly chlorinated pesticides and fungicides, but also transformer coolants (PCBs) and combustion by-products (polychlorinated dibenzofurans and dibenzodioxins). A further 14 chemicals have been added to the list since ratification, including flame retardants (e.g. hexabromobiphenyl, hexabromocyclododecane) and so-called durable water repellents for textiles (perfluorooctanesulfonic acid).

The Convention allows for some products containing banned chemicals which were produced prior to their regulation to be used until the end of their lives. It also allows certain exemptions in cases where the banned substances may be necessary. For example, although DDT is a POP, has caused major environmental impacts and is one of the original dirty dozen, the Convention allows for it to be used in compliance with World Health Organization (WHO) rules for safe disease vector control (i.e. mosquito hazard management) (WHO, 2004).

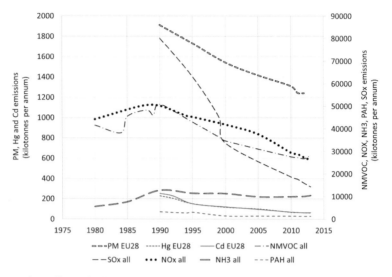

Figure 8.2 Reduction of air pollutants by CLRTAP signatories. PM – particulate matter; NMVOC – non-methane volatile organic carbon; PAH – polyaromatic hydrocarbons (data: UNECE, 2017)

8.1.2 CLRTAP

The Convention on Long-Range Transboundary Air Pollution (CLRTAP) is a product of a sub-group within the UN – the United Nations Economic Commission for Europe (UNECE), which, despite its name, also includes the United States, Turkey and several Central Asian republics (51 countries altogether). It was signed in 1979 and entered into force in 1983. As signatories to this over-arching convention and its more detailed protocols, these countries have committed to reduce the emission of airborne pollutants such as oxides of sulphur and nitrogen, heavy metals, volatile organic carbon (VOC) compounds and POPs. The most recent example of a protocol under CLRTAP is the Gothenburg Protocol, which in 1999 set national emission ceilings for SO_2, NO_x, VOCs and NH_3 for the period 2010–2020. An indication of the success of this UNECE agreement is presented in Figure 8.2, which shows the total amount of regulated emissions among signatories ('all') or among 28 European states ('EU28') falling under the relevant periods of the Convention.

Particularly interesting features of the CLRTAP include the *critical loads approach* and the *multi-pollutant, multi-effect approach* to pollution. The critical loads approach puts focus on the extent to which different geographical areas can tolerate different pollutant deposition loads. The multi-pollutant, multi-effect approach focusses on all the different kinds of effects that all the various pollutants can give rise to in an integrated manner. These approaches to regulation are made possible by the broad coverage of air pollutants of the CLRTAP and they allow for emission targets to be tailored to the specific needs of both emitting and receiving countries.

8.1.3 Montreal Protocol

The Montreal Protocol on Substances that Deplete the Ozone Layer sets binding targets for the reduction of the emission of ozone-depleting gases under the more general 1985 Vienna Convention for the Protection of the Ozone Layer. It was signed in 1987 and came into force in 1989. It is perhaps the most successful international environmental treaty ever. It has been ratified by all the members of the UN and several other parties, including the EU and the Holy See (Vatican).

The Protocol identified five key chlorofluorocarbon (CFC) gases ($CFCl_3$ – 'CFC-11', CF_2Cl_2 'CFC-12', $C_2F_3Cl_3$ 'CFC-113', $C_2F_4Cl_2$ 'CFC-114' and C_2F_5Cl 'CFC-115') which were phased out between 1989 and 1996.

CFCs have been widely used as refrigerator gases, as propellants for aerosols and as solvents (see Section 3.5 for a discussion of their environmental effects). In many cases, hydrochlorofluorocarbons (HCFCs) were used as interim substitutes to allow the phase-out of CFCs. The HCFCs generally had lower potential to deplete the ozone layer, and also had lower global warming potential. HCFCs are to be phased out over the period 2015–2030. An amendment to the Montreal Protocol to additionally phase out hydrofluorocarbons (HFCs) was signed in 2016 in Kigali, Rwanda, and is expected to come into force in 2019 after ratification by 20 countries. It aims to eliminate 85% of HFCs by 2040. The HFCs have the advantage of having no potential to damage the ozone layer, but their greenhouse gas characteristics are problematic – they are up to 10,000 times worse than carbon dioxide on a mass basis.

Industrial interests fought a long battle to protect existing markets and to hinder policy makers from implementing the Protocol. As late as 1988, Richard Heckert, then the chairman and CEO of DuPont, wrote to the US Senate saying: 'There is no available measure of the contribution of CFCs to any observed ozone change.' Recent atmospheric research has shown that since the treaty was signed, and despite some illegal CFC emissions, the ozone layer is recovering: the concentration of ozone is increasing and the scale of the Antarctic ozone hole is diminishing (Solomon et al., 2016). According to the US Environmental Protection Agency (USEPA), by protecting the ultraviolet radiation filter which the ozone layer provides (see Section 2.1.1), the Montreal Protocol will prevent 283 million skin cancers, 46 million cataracts and 1.6 million deaths among citizens of the United States. Globally, the UNEP expects 2 million skin cancers to be prevented each year by 2030 (UNEP, 2015); see Figure 8.3. There are a number of reasons for the notable success of the Montreal Protocol. For example: the relevant industrial emissions, primarily the CFCs and halons, had point sources rather than diffuse sources; there were practical substitutes to these chemicals; the effect of a thinner ozone layer was understandable and frightening for most people, and the science around stratospheric ozone depletion was not debated much.

8.1.4 Kyoto Protocol and Paris Agreement

The Kyoto Protocol and the Paris Agreement are two of the more famous elements of the United Nations Framework Convention on Climate Change (UNFCCC). This

Predicted abundance
Thousand parts per trillion

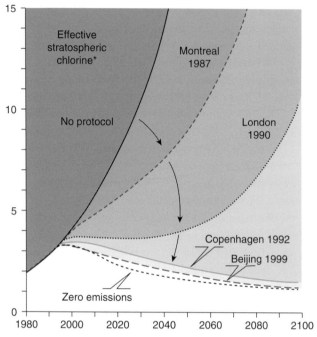

* Chlorine and bromine are the molecules responsible for ozone depletion.
'Effective chlorine' is a way to measure the destructive potential of all ODS
gases emitted in the stratosphere.

Cases per million people per year

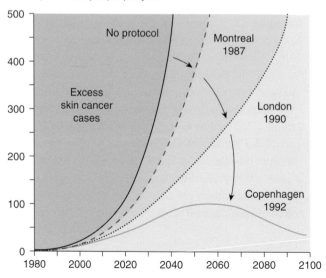

Figure 8.3   Effective chlorine loading and predicted excess skin cancer cases after regulatory protocols (Emmanuelle Bournay, GRID-Arendal, http://grid-arendal.herokuapp.com/resources/7524, reproduced with permission)

Convention was one of the products of the 1992 Earth Summit in Rio de Janeiro, Brazil, and came into force in 1994. The UNFCCC did not itself contain detailed targets or compliance mechanisms to fight climate change, which is why the agreements reached in Kyoto and Paris are important.

The Kyoto Protocol was agreed in 1997 and came into force in 2005 after ratification by 55 member states. The Protocol established a 'commitment period' between 2008 and 2012 during which parties to the Protocol were required to reduce their emissions by negotiated amounts (typically about 5%) compared with 1990 levels. Only wealthier, industrialised countries (called 'Annex B' countries in the Protocol) were asked to reduce emissions. There were some exceptions – for example Australia, which is a wealthy country that had a conservative government at that time, and negotiated to limit itself to an 8% *increase* in emissions. Countries were required to report their greenhouse gas emissions annually. A 'Clean Development Mechanism' and 'Joint Implementation' mechanism were created to allow wealthy countries to earn emission credits cheaply by investing in poorer countries or each other (respectively). An international greenhouse gas emissions trading mechanism was also set up. Enforcement provisions for the Protocol were not strong compared with domestic law, but included that (1) a country in breach of its commitments would be required to reach them in the second commitment period, plus an additional 30%; and (2) the country would be suspended from the trading mechanism.

Several important developments have occurred since the end of the first commitment period. A 'Green Climate Fund' was formally established in 2010 under the UNFCCC to support investments around the world. It began making investments in 2015. Also, in 2012, a second commitment period under the Kyoto Protocol (2013–2020) was negotiated via the Doha Amendment to the Protocol, with a slightly different group of nations (e.g. Canada withdrew). By October 2018, 120 nations had ratified the Doha Amendment, which requires 144 states to ratify it in order to come into force. 36 nations ratified it during 2018, but there is a risk the Doha Amendment may not come into force before the expiration of its commitment period. This situation may in part be due to the parallel development of the Paris Agreement.

The Paris Agreement is a separate document under the UNFCCC which was negotiated in 2015. One hundred ninety-seven parties (mostly countries plus, for example, the EU) have signed the agreement. Enough countries ratified it by November 2016 to cover 55% of global emissions, bringing it into effect.

The aim of the Paris Agreement is to hold human-induced climate change below 2°C and preferably limit it to 1.5°C. Individual countries get to decide the scale of their emission reduction commitments. Unlike the Kyoto Protocol, there is no differentiation between rich and poor countries, although it is understood that poorer countries may have different capacities to reduce emissions. Also unlike the Protocol, there is no enforcement mechanism other than 'naming and shaming' countries that do not live up to their promises. Nevertheless, consistent with the Kyoto Protocol, reporting national emissions to the UN is mandatory. An informally named 'Sustainable Development Mechanism' was created as a successor to the 'Clean Development Mechanism' of the Kyoto Protocol. All countries can participate in the new Mechanism. The Green Climate Fund is also intended to continue to support climate mitigation efforts towards meeting the aim of the Agreement.

Regulating greenhouse gas emissions is fundamentally more difficult than regulating ozone-depleting substances. Overall, the Paris Agreement seems to be an improvement on the Kyoto Protocol, but both have faced intense opposition from wealthy interests. At the time of the writing of this book, American President Donald Trump had announced he will extract the United States from the Agreement, claiming it is bad for business and for less affluent Americans. Under American law, it is considered an 'executive agreement' made by the previous president, Barack Obama, rather than a 'treaty' made by a two-thirds majority of the US Senate, so it requires little effort for the president to unmake his predecessor's commitment to the Agreement. (Despite resistance in the US Senate, Obama was legally able to commit the United States to it because the Senate had previously ratified the UNFCCC.) The global response to date has effectively been to ignore Trump – since the United States now accounts for only 14% of global emissions and there is widespread global endorsement of the Agreement, he cannot directly end it. Some of the subsequent political news is that even poor, war-torn Syria has now agreed to ratify it.

8.1.5 Basel Convention

The full name of this United Nations treaty is the Basel Convention on the Control of Transboundary Movements of Hazardous Wastes and Their Disposal. Its overall aims are to prevent toxic waste being dumped by wealthy countries in poorer countries, to reduce the amount and toxicity of wastes being generated and to help poorer countries manage their own wastes better. This became a pressing need when various pollution incidents came to light in the 1980s. For example, the ship *Khian Sea* dumped 3,630 tonnes of American incinerator ash on a beach in Haiti as 'topsoil fertiliser' before Greenpeace informed the Haitian government what the material actually was. The ship then toured the world, changing its name several times and being rejected in numerous ports before dumping its remaining 907 tonnes at sea. This incident and others showed that because waste management was becoming safer and more expensive in wealthy countries, and the cost of shipping was falling, there was a serious new problem to solve. The Basel Convention was signed in 1989 and came into force in 1992. One hundred eighty-four countries and the EU as a whole have ratified the convention. Haiti and the United States signed it but have not ratified it.

The Convention's definition of hazardous waste includes waste that is considered infectious, flammable, toxic, corrosive or explosive. Examples of these include biomedical wastes, used oil, used lead-acid batteries and POPs. The Convention places requirements on countries to collect and manage information about international waste transportation. Parties to the Convention (countries) are required to prevent exports of hazardous waste by private operators to other countries which have banned the import of hazardous waste. Even if the recipient country has not banned such imports, transport of hazardous waste by waste management companies between parties to the Convention is illegal without prior consent of those parties.

Countries have to pass legislation punishing illegal waste transportation. There is an obligation to reduce the amount of hazardous waste being generated, and to treat it close to its source. The Convention negotiations also established an international network of 'Regional and Coordinating Centres' which provide training in hazardous waste management.

8.2 The European Union: A Multinational Jurisdiction

The EU is a significant political institution. At the time of writing, the EU had 28 member states and a population of more than 510 million people. The government of the EU has four key institutions reflecting its organic origins as a steel cartel that turned into a democratic institution. The European Council consists of the heads of state of each member of the EU. It cannot pass law, but it sets the EU's management agenda. It is backed up by the Council of the European Union representing national governments, but its legislative powers are balanced by the European Parliament, which is directly elected by the citizens of the EU, and the European Commission, which has about 23,000 staff members who develop legislation and, if approved, enforce it. Depending on the type of legislative instrument being considered and its topic, consent by both the Council of the European Union and the European Parliament, or only one of these two, may be required.

Overall, the EU is a unique political organisation somewhere between a confederation (in which central government legislation typically has to be enacted by member states in order for it to be effective) and a federation (where state governments have unique powers, but the central government can act on certain matters without referring to state parliaments). Despite this ambiguity and the complexity of its institutional structure, the EU has played an important role in environmental regulation in Europe.

The EU has lobbied for and is a signatory to many of the international environmental laws listed in the previous section. The EU creates three kinds of legislative instruments that can govern various aspects of environmental management within EU member states: *regulations, directives* and *decisions*. The Council of the European Union and the European Parliament can make regulations. These apply to the whole Union from the moment they are passed. Some key examples of this in the context of this book are the REACH regulation, the Euro 6 regulation (see later in this chapter) and the CLP regulation ('Classification, Labelling and Packaging') which implements the Globally Harmonised System of Classification of Chemicals (see Chapter 6). *Directives* require member states to reach certain goals but leave determination of the form of implementation to the national parliaments. Examples of some key directives relevant to this book are the RoHS Directive and the Water Framework Directive (both are described later in this chapter). *Decisions* can be taken by the European Commission without referral to the European Parliament, and tend to have a smaller scope than *regulations* or *directives* – so they are not discussed further here.

8.2.1 The REACH Regulation

The REACH ('Registration, Evaluation and Authorisation of Chemicals') regulation is the EU's key regulation for managing chemical hazards. It was passed in 2006 and is administered by the European Chemicals Agency (EChA), headquartered in Helsinki, Finland, though agencies of European national governments are responsible for monitoring compliance. Article 57 of REACH defines 'a substance of very high concern' (SVHC) as one of the following:

- carcinogenic
- mutagenic
- toxic to reproduction
- persistent, bioaccumulative and toxic (PBT)
- very persistent and very bioaccumulative (vPvB)
- a substance of equivalent concern

The first three items on the list are sometimes grouped with the abbreviation 'CMR', but the criteria are independent. On the other hand, in the case of PBT (and vPvB) substances, each of the three criteria (or two criteria for vPvB) has to be fulfilled for the substance to be classed as an SVHC. The last relatively general phrase on the list is designed to catch substances like endocrine disruptors, respiratory sensitizers and other currently unknown substances that may need regulating.

The REACH regulation requires companies that wish to manufacture or import a chemical into the EU in a volume of 1 tonne per year or more to register it with EChA. In other words, 'no data, no market'. If the chemical is to be marketed in the EU at a rate of more than 10 tonnes per year, a Chemical Safety Report (a hazard assessment) is required, which describes its physiochemical properties, health and environmental characteristics and whether it is PBT or vPvB. If it is identified as hazardous, an exposure assessment and risk characterisation is required for all known exposure pathways.

After registration, EChA performs an evaluation based on the submitted data. There are two kinds of evaluations: *dossier evaluations* and *substance evaluations.* *Dossier evaluations* are compliance checks in which EChA examines the material submitted by companies when they register substances; EChA examines 5% of dossiers in this way. About half are considered incomplete and are returned to the companies for improvement. In cases where it is suspected that the chemical may be an SVHC, EChA collates all the dossiers for the chemical and performs a more extensive *substance evaluation* in collaboration with a member state (i.e. a European country). This may involve requesting companies to provide new toxicological or environmental monitoring data.

Under REACH, 'authorisation' is the opposite of what some students imagine – a chemical substance that has been determined to be an SVHC is generally banned, but authorisation for particular use can be made on a case-by-case basis. The list of these banned chemicals is called Annex XIV (the 'Authorisation List'). This list is short but growing – it had only 9 entries in 2011 and 43 at the end of 2018. For people concerned about the need to regulate many more of the 14,000 chemicals registered under REACH, this may seem like a modest result. But there are other levels of regulatory influence under REACH. One of them is the 'Candidate List', which is EChA's list of chemicals under consideration for placement on Annex XIV. At the time of writing, 228 chemicals or classes of chemicals were on this list (EChA, 2017). From the moment a chemical is placed on the Candidate List, companies are obliged to provide chemical safety information to customers, even if the chemical is only present as more than 0.1% of the mass of a product. For example, if a certain chemical on the Candidate List makes up 0.2% of a shoe, and a customer in a shoe shop asks for this safety information, the retailer must provide it within 45 days, or break the law. Consequently, inclusion of a chemical in the Candidate List is enough of an incentive to get many companies to urgently look for substitutes.

Another key element of REACH is the 'Restriction List' – Annex XVII. There were 533 chemicals or classes of chemicals on the list in 2017 – 211 of which were some kind of cadmium compound. Substances in Annex XVII can be used by manufacturers and other companies, but they are subject to some kind of specific use restrictions. For example, nickel may not be used in earrings unless the leach rate is less than 0.2 $\mu g/cm^2$/week.

8.2.2 RoHS Directive

The 'Directive on the Restriction of the Use of Certain Hazardous Substances in Electrical and Electronic Equipment 2002/95/EC', unsurprisingly abbreviated to 'Restriction of Hazardous Substances Directive' or simply RoHS, is a predecessor to REACH which took effect in 2006 and is still in force. It restricts the 10 substances shown in Table 8.1 within household electrical and electronic equipment.

Any single homogeneous material in such equipment (for example, the insulating sheath on a cable is considered to be one) may not have a concentration of any of the substances listed in the table which exceeds 0.1% of the total mass. For cadmium, the limit value is even lower: 0.01%. There are some significant exceptions to the RoHS directive, for example batteries, but these are covered by another directive (the Battery Directive 2006/66/EC). There is no specific labelling scheme for goods

Table 8.1 Substances Regulated by RoHS

Lead

Mercury

Cadmium

Hexavalent chromium

Polybrominated biphenyls

Polybrominated diphenyl ether

Bis(2-ethylhexyl) phthalate

Butyl benzyl phthalate

Dibutyl phthalate

Diisobutyl phthalate

Figure 8.4 European compliance label

claiming compliance with the RoHS directive, but the more general European Compliance label (which relates to a host of European directives) is used (see Figure 8.4).

8.2.3 The Euro 6 and RDE Regulations

Several regulations, directives and decisions of the European Union have relevance to the regulation of air quality. Two related examples are Regulation 2012/459, which promulgated the 'Euro 6' standard for emissions from light passenger and commercial vehicles, and Regulation 427/2015, which tightened it. Euro 6 requires that cars emit less than 0.06 or 0.08 g NO_x per kilometre for petrol or diesel engines, respectively. Even before the 'Dieselgate' scandal (the discovery that some car manufacturers had programmed cars to cheat emissions control tests) became public in 2015, it had been apparent for some years that NO_x emissions to air from cars were not falling in Europe as fast as had been expected, since real road conditions are different to laboratory test conditions. Regulation 2016/427 introduced a 'Real Driving Emissions' provision by which cars are tested outside laboratories using mobile measurement technology, to ensure they comply with the standards. It is hoped that legislative efforts such as these will improve air quality, particularly in major cities where air pollution causes premature deaths.

8.2.4 Water Framework Directive

The Water Framework Directive (officially called 'Directive 2000/60/EC of the European Parliament and of the Council of 23 October 2000 Establishing a Framework for Community Action in the Field of Water Policy') is an example of legislation that has to be interpreted by national governments before implementation. It is a management framework intended to help Europe achieve better freshwater aquatic environments. It also relates to the first nautical mile of the offshore marine environment. Water bodies are classified under the Directive as good or poor according to their biological, chemical and physical conditions. For example, if the maximum stipulated concentration of any individual chemical pollutant is exceeded in a river, it cannot be classed as having 'good ecological status'. Hydromorphological aspects such as river bank conditions are also considered. Critically in the EU context, the Directive instigates the creation of River Basin Management Plans based on hydrological catchments, rather than national political boundaries. This element is intended to force collaboration between member states to improve river management.

The Water Framework Directive has been relevant to water resources planning. For example, there was a plan to divert the Ebro River in northern Spain to promote economic growth on the dry south-eastern Spanish coastline. Among other concerns, the plan was seen to have direct hydrological and ecological impacts on the Ebro delta, a Ramsar-listed[1] wetland, and considered to conflict with the Framework Directive, so it was eventually cancelled (Tortajada, 2006). Instead, plans were made for the construction of 20 seawater desalination plants, along with water recycling facilities and more efficient agricultural practices (ICEX, 2009).

8.3 National Jurisdictions

The United States of America and the Kingdom of Sweden are both jurisdictions governed via representative democracy, meaning that environmental laws have to be passed by at least one chamber of representatives elected by the citizenry. Although they have this characteristic in common, they are very different in other ways, making them useful examples of different kinds of national jurisdictions. So this chapter focusses on these two jurisdictions as a way of showing how lawmaking at the national level differs between nations, and how it interacts with other levels, including (in the case of Sweden) the EU.

8.3.1 United States

8.3.1.1 How Laws Are Made in the United States

The United States is a federation of states, meaning that a significant number of powers are devolved from the federal government to the states. The national legislature of the United States is called the 'Congress' and is bicameral (consists of two chambers): the House of Representatives and the Senate. The House of Representatives is intended to reflect the principle of proportional representation, so populous California has 53 representatives, while wild Alaska has only 1. On the other hand, the Senate has two representatives per state, regardless of the states' population. Naturally, this can lead to tension between the chambers, but this organisational principle is not uncommon globally (e.g. the Australian parliament has a similar bicameral arrangement), and it has been necessary in some cases to ensure that less populous states will join the formation of a nation. To become law in the United States, a bill (a draft version of the law) has to be passed by both chambers of Congress and then signed by the president. A president can veto a bill or send it back to Congress with improvement suggestions, but if both chambers of Congress pass it with a two-thirds majority, the president can be overruled. Congress and the presidency have historically been dominated by two political organisations: the Republican Party and the Democratic Party.

---

[1] I.e. important enough to warrant listing for protection under the United Nations Educational, Scientific and Cultural Organization's 'Ramsar Convention on Wetlands of International Importance Especially as Waterfowl Habitat'. Ramsar is the city in Iran where this convention was signed in 1971.

Table 8.2 Types of Legal Instruments around the World		
Type of statute	Swedish translation	Responsibility
Law	*Lag*	Being general and wide-ranging, laws must be voted on by a legislature (parliament).
Regulation	*Förordning*	A government can write these. It may also delegate to a public authority (e.g. USEPA).
Directive, Guide, Code of Practice	*Föreskrift*	Specific and detailed, these can only be made by a specialised regulatory authority (e.g. an EPA).

As with the foundational documents of many nations, the constitution of the United States does not directly address environmental matters. Instead, the ability of the national government to create environmental law rests on its power to regulate interstate commerce. This power exists in the constitution to prevent a state using legislation to serve its interests in a way that negatively affects other states. Many environmental laws and regulations have practical consequences for interstate trade, and can therefore be defended from accusations of being unconstitutional. The national government also has jurisdiction over federal lands (mostly arid and grazing land in the west) and national parks. This is no small matter, as about 28% of the surface area of the United States is classified as federal land (about 60% of this is in Alaska). Congress passed several major environmental laws in the second half of the past century, which have been maintained and updated since then. Examples include: the National Environmental Policy Act, 1969; the Clean Air Act, 1970; the Clean Water Act, 1977; and the Superfund Act, 1980.

In 1970, Congress also approved President Nixon's proposal to centralise national environmental management functions by creating the USEPA. The USEPA is a large government institution which had more than 15,000 employees and a budget of $8 billion in 2016. This scale has been necessary because of the USEPA's nationwide responsibilities for environmental monitoring and administration of the laws mentioned earlier. The content of some key environmental laws is summarised in the next section. These laws do not provide enough detail in most cases, so the USEPA has an important role in interpreting these laws and making practical decisions about how to apply them (Antweiler, 2014). The USEPA has the power to make detailed regulations that supplement environmental laws and make them practical enough to be implemented (see Table 8.2). When the USEPA proposes a regulation, the text is exposed to the public via listing in the *Federal Register*, an official, daily online publication for enacted and proposed rules and regulations (see www.gpo.gov). It is then revised and published in the Code of Federal Regulations, typically in the volume called Title 40. (Both the *Register* and the Code can be accessed via www.gpo.gov/fdsys/search/home.action.)

8.3.1.2 Key Environmental Laws

The National Environmental Policy Act (NEPA) 1969 established a requirement that actions by federal agencies which may have a significant impact on the environment must be first the subject of an environmental assessment (EA). Actions by states which are not connected with federal funding are exempt. An EA is brief compared with an environmental impact assessment (EIA). There is typically a 15-day public exposure period after which the agency that produced the EA has to decide whether an EIA is warranted. EIAs are described in Section 8.4 of this chapter.

The Clean Air Act of 1970 was not the first law with that name, but it was the first 'with teeth' in the sense that it instituted comprehensive state and federal regulations and enforcement provisions. The Act required the USEPA to define quantitative national air quality criteria for a list of hazardous air pollutants. Since Congress gave the Obama administration little legislative freedom to regulate greenhouse gas emissions via potentially more efficient market mechanisms, six greenhouse gases were listed as hazardous air pollutants in 2009 as a way to apply a more traditional regulatory lever to the climate change problem. The Act requires states to develop regulatory plans to reach the air quality criteria for listed pollutants. Both stationary point sources and mobile sources (e.g. vehicles) are considered. The USEPA monitors air quality around the country to determine whether the criteria are being exceeded. It has the right to intervene in state affairs when criteria are exceeded, by applying specific requirements on particular industrial facilities, but generally prefers to collaborate with the states.

The Clean Water Act is an example of an environmental bill which the president (Nixon) vetoed but which nevertheless passed into law. The Act provides a basis for the licensing of point sources of aquatic pollution. The licences contain quantitative limits, for example an effluent may be required to have a biochemical oxygen demand (BOD) of less than 25 mg/L. The Act also allows the USEPA to issue administrative orders and to prosecute unlicensed polluters or those that exceed their licence conditions. In cases of criminal negligence, a violator can be fined $2,500–$25,000 per day and/or spend a year in jail. The fine doubles for a second offence. Where the crime is deliberate rather than negligent, and places others at serious risk of injury, an individual can be fined $250,000 and/or face 15 years in jail. An organisation can be fined $1 million.

The Comprehensive Environmental Response, Compensation, and Liability Act of 1980 is also known as the Superfund Act, after the name of the pot of money it collected to deal with contaminated land problems. The most important of these was Love Canal, a toxic waste dump near Niagara Falls which had been converted into a residential area, and which became a media event in the 1970s on account of the obvious and serious health effects on residents. The Act authorises urgent, relatively simple 'removal actions' to deal with urgent threats, and 'remedial actions' to clean up contamination on site. After the USEPA is notified of the existence of some contaminated land, samples are taken and a standardised *hazard ranking system* is used to give the site a rating between 0 and 100. This enables nationwide prioritisation of the sites that pose the worst threats to people and the environment, based on their hazard and exposure characteristics. If the site scores 28.5 or more, it is eligible for clean-up funding from the superfund, but at the majority of sites, remedial activity is funded by the current owner or its previous users, when their culpability can be demonstrated. The superfund is available for the 30% or so of sites where no owners or users can be tapped for the cost of remediation. The superfund was originally recharged by a tax on the oil and chemicals industries, in accordance with the polluter pays principle, but it is now funded by US taxpayers.

A new key law for the regulation of chemicals in the United States is the Frank R. Lautenberg Chemical Safety for the 21st Century Act (FRL Act) of 2016. Unusually for this period of government, the law was passed with large bipartisan margins in both the House and the Senate. Its role is to update the Toxic Substances Control Act (TSCA) of 1976. It provides greater regulatory oversight, including:

- A duty for the USEPA to evaluate chemicals currently in use, with enforceable deadlines
- A basis in risk assessment rather than the risk-benefit analysis used in TSCA
- Powers to more rapidly demand information from companies about chemicals
- New chemicals or new uses of old chemicals must be approved before use

In these respects, the act is similar in intent to the EU's REACH legislation. The USEPA can declare that a new chemical 'presents an unreasonable risk', that a new chemical 'is not likely to present an unreasonable risk', or that information is insufficient for the evaluation. While the subsequent choice of actions to manage risks should be based in part on their relative cost, the classification of chemicals is purely risk based. Existing chemicals are prioritised on the basis of their threat to people or the environment, to ensure that the most important ones are assessed soon – 'high-priority' chemicals must be evaluated within 3.5 years. The FRL Act also provides the USEPA with a basis to make manufacturers pay higher fees than previously to help to fund this regulatory oversight. For some chemicals, the USEPA can charge the manufacturer the whole cost of regulatory review, up to a total annual programme cap of $25 million.

8.3.1.3 The States Versus the United States

As was previously mentioned, Congress' ability to regulate interstate trade exists in part to ensure that states do not pass protectionist legislation that disadvantages other states. This intent is one of the reasons why, despite the nation's federal political structure, environmental regulation in the United States is dominated by the national scene. Historically, the states have not been entitled to fill in regulatory gaps left between federal laws. Instead, states typically regulate to the extent that federal law allows, and use the same environmental quality standards. However, there are some notable exceptions to this. One example is California's exemption from elements of the Clean Air Act. California already had emission standards for motor vehicles before this Act, and was given permission to enforce more stringent standards than the national standards. Subsequently, some other states have elected to apply the Californian standards. In the courts, California has also successfully defended market interventions to promote zero-emission vehicles.

A parallel to California's partial Clean Air Act exemption exists under the Clean Water Act. It says that states are allowed to promulgate their own water quality standards for particular lakes and rivers within a state, as long as the standards are tougher than the general federal ones. Likewise, the FRL Act is national, but it preserves the rights of state governments to regulate in ways and for chemicals that the USEPA does not choose to regulate.

More generally, the standards laid out in the Clean Air Act and the Clean Water Act are national, but states get to decide how the standards are to be reached – whether through market mechanisms, investments or innovations in licencing practices and enforcement. Sixteen states also have their own EIA legislation, for example the California Environmental Quality Act 1970, which regulates public and private development projects in that state when they are not covered by the NEPA 1969. However, in 13 of these 16 states, private developments are exempt from the requirement to produce an environmental impact statement (EIS – see Section 8.4).

8.3.1.4 The Judiciary

Lawsuits have played an important role in environmental regulation in the United States. The United States has many courts, at both the state and federal levels. Because the United States is a federation, the naming conventions and number of layers in the state court hierarchy are variable, but it is common to have a three-step hierarchy including a state supreme court, circuit courts and district courts. The federal court system has a supreme court, courts of appeal and district courts, separate from the state district courts. There are also many specialised courts dealing with particular topics like tax or immigration. Environmental lawsuits are initially heard in the court that is relevant to the particular law, and may progress all the way to the US Supreme Court if there is some matter of interpretation of federal law that requires adjudication. Most federal laws allow individuals and organisations to sue other individuals who have misapplied or broken the law, including individuals working in governments and the USEPA. The main requirements are that the plaintiff can demonstrate harm to themselves, and that the harm is ongoing rather than historical (Antweiler, 2014).

One example of such litigation was the case of *Sierra Club* v. *Babbitt* in 1998. The Sierra Club is a non-governmental, outdoor recreation and nature conservation organisation. Bruce Babbitt was the US secretary for the interior, responsible for the US Fish and Wildlife Service (FWS). The Sierra Club alleged that the FWS had erroneously issued a permit for the construction of high-density housing in the habitat of an endangered mouse species. The relevant US district court agreed and determined that the permit was not consistent with the NEPA 1969, forcing the construction project to halt.

Another notable case was *Entergy* v. *Riverkeeper* in 2009. Entergy is a power company using river water as a coolant. Riverkeeper is a non-governmental environmental conservation organisation focussed on the Hudson River. Riverkeeper wanted stronger protection for the river and objected to the impacts on aquatic life caused by the cooling water intakes. The US Supreme Court decided that it was acceptable for the USEPA to consider the results of a cost-benefit analysis in the process for identifying the best available technique (see Section 1.3.4.3) for power station cooling systems. This was a landmark court case because most federal environmental laws do not support, or in fact actively prevent, the use of such financial considerations in decision-making. The court agreed with Entergy, which set a precedent for the increased use of cost-benefit analysis in environmental management by the USEPA. This court intervention in USEPA procedures may be expected to weaken the level of environmental protection over the medium term.

Unfortunately, while in principle the court system offers some hope against injustice, it frequently tends to favour commercial interests over environmental ones. One of the fundamental problems is the cost of going to court. While individuals concerned about pollution typically have to find or create a non-governmental organisation and collect funds from supporters in order to approach a court, a large private company may have a legal department and staff with appropriate courtroom skills already on its payroll. Additionally, a private company can report the expense of running a court case as a cost of doing business, thus subsidising its court expenses via a reduction in corporate tax. This is a privilege not extended to ordinary citizens and non-profit organisations.

8.3.2 Sweden

8.3.2.1 How Laws Are Made in Sweden

As a constitutional monarchy, Sweden does not have a president as head of state and instead has a monarch with ceremonial functions. Sweden's legislative powers rest with a unicameral parliament (Riksdag) elected by proportional representation, like the US House of Representatives. Unlike the US Congress, the Riksdag has for the past 50 years had representatives from five or more political parties. In 2018, the organisations included, in descending order of representation: the Social Democratic Party, the Moderate Party, the Swedish Democrats, the Center Party, the Left Party, the Christian Democratic Party, the Liberal People's Party and the Green Party. Sweden has also elected representatives of other political parties to the European Parliament, including delegates from the Pirate Party[2] and the Feminist Initiative.

The prime minister of Sweden usually represents the largest party in the Riksdag and manages the government, usually based on an agreement with parties that make up a simple majority of the parliament, but the prime minister has only one vote and lacks the veto power which the president of the United States has. Laws are proposed by the government or other parties in parliament, and require a majority vote in the Riksdag to come into force. As mentioned previously, EU regulations become law in Sweden when they are activated in Brussels, but EU directives require national legislation for implementation. For example, when the EU made the Water Framework Directive, the Riksdag voted through a change to chapter 5 of the Swedish Environmental Code, which now describes the geographic extent of five catchments for the River Basin Management Plans required under the Directive.

Sweden is not a federation but a unitary state, which means more power is kept at the national level. The Swedish level of government equivalent to a state in the United States is a county (*län*). Counties primarily administer health and transportation policy, but also have some environmental responsibilities. Since the national

---

[2] This party is not focussed on armed theft at sea, but primarily on combating corporate or government spying on citizens using electronic media, the excesses of patent law and other issues connected with personal liberty in the information economy.

parliament has only one chamber (i.e. the Riksdag), there are no tensions between competing chambers of national government and with the presidency that offer quality control for new legislation and some of the limits to executive power in the United States. On the other hand, with so many parties present in the Riksdag, the largest (Social Democrats) have only achieved an absolute majority twice in the past 100 years, so the Riksdag has been able to offer more room for negotiation and debate than many other legislatures around the world, thus compensating for its unicameral structure.

8.3.2.2 Key Environmental Laws

Three kinds of statutes (*författningar*) are recognised in Swedish law (see Table 8.2). In 1999, a new environmental law was enacted by the Swedish parliament that was intended to simplify environmental management by abolishing 15 other intersecting laws and uniting their common elements in a single law. Known as the Environmental Code (*Miljöbalken*), it covers pollution of the air, water and soil, the protection of human health and the natural environment and the management and recycling of natural resources.

The Environmental Code lists a set of general rules (*Allmänna hänsynsregler*) which may seem vague at first but in fact are used to justify the decisions of licensing authorities, and thus are connected to more detailed regulations and individual decisions. These rules are summarised next:

1. Burden-of-proof rule – it is up to the proponent of an action to prove that it is safe, not up to the government to prove the action is dangerous.
2. Knowledge rule – the proponent must acquire the knowledge needed to protect human health and the environment from the action.
3. Precautionary rule – the proponent must use the best available technique to prevent any potential harm, rather than waiting for proof that the harm can occur.
4. Product choice rule – the user of chemical or biological materials must choose the least damaging materials.
5. Efficiency rule – users of energy and materials should seek to use them efficiently, to use renewable energy and to recycle materials.
6. Localisation rule – activities should be located where they are most environmentally appropriate, consistent with local area plans.
7. Reasonableness rule – the other rules should not be applied to excess; costs and any military needs should be considered.
8. Remediation rule – if you pollute, you are responsible for cleaning up.
9. Prohibition rule – only the government can allow an activity that damages the environment. If the activity will compromise human health, not even the government can permit it.

For any action that may impact the environment, the Code reverses the burden of proof – the proponent of the action has to show that it complies with these rules.

Chapter 5 of the Code authorises the government regulatory authority to determine environmental quality standards, and, by itself or in dialogue with lower levels of government, to create intervention programmes to ensure those standards are reached.

As shown in Table 8.2, regulations provide more detailed information than laws and can be changed more easily (by the relevant ministry) without having to go back to the Riksdag for endorsement. An example of this is regulation SFS 2013:251 of the Swedish Ministry of the Environment and Energy. Ninety per cent of this 42-page document is a list of potentially environmentally hazardous economic activities. It indicates which government organisation is authorised to assess the environmental impact statement and make a decision to approve the activity. Activities are classified 'A' – to be assessed by one of the specialised Land and Environment Courts (see Section 8.3.2.3); 'B' – to be assessed by the Country Environmental Permit Board (*Miljöprövningsdelegationen*); or 'C' – to be assessed by the local municipal Environmental Health Committee (*Miljö- och hälsoskyddsnämnd*). For example, a waste recycling facility handling more than 100,000 tonnes of waste per year is classified 'A'; otherwise it is classified 'B' for values above 500 tonnes per year, or 'C' at or below that value.

Directives are created by organisations with direct responsibility for environmental management, and do not necessarily have to be directly endorsed by the relevant minister. For example, the Swedish EPA has directive NFS 2016:7, which describes the kinds of information which holders of environmental pollution licences must provide in their annual reports to the Swedish EPA.

8.3.2.3 The Role of the Judiciary

The judicial system in Sweden is actually two parallel systems with three tiers each. The general courts (*allmänna domstolar*) hear criminal and civil cases, while the general administrative courts (*allmänna förvaltningsdomstolar*) hear disputes between individuals and government agencies. In the first tier, there are 48 general district courts (*tingsrätt*) and twelve administrative courts (*förvaltningsrätt*). In the second tier, there are six general appellate courts (*hovrätt*) and four administrative courts of appeal (*kammarrätt*). Finally, in the third tier, there is a supreme court for general matters (*högsta domstolen*) and another for administrative matters (*högsta förvaltningsdomstolen*). There are also specialised courts, including the land and environment courts (*mark- och miljödomstolar*), five of which exist at the district level, with the potential to appeal to an appellate court in Stockholm (*Svea hovrätt*). General district courts can hear cases based on breaches of environmental law, but the land and environment courts only hear cases relating to:

• Environmental permits for environmentally hazardous actions
• Compensation claims for environmental problems
• Appeals against land use planning decisions

Historically, Sweden has relied less on the court system for decision-making than the United States. Sweden has a strong tradition of 'civil' (or statutory) law that can be contrasted with the 'common law' tradition of countries like the United States and the United Kingdom. That is to say, in Sweden, the laws passed by the government play a relatively significant role, whereas in jurisdictions where 'common law' is more important, the accumulated history of decisions by the court system provides more guidance than it does in Sweden. Consequently, it has been suggested that Swedes are less likely than Americans to seek redress through the courts, favouring mediation instead (Ortwein, 2003).

8.3.2.4 Sweden's Environmental Objectives

As is the case in the United States, Sweden has centralised many of the environmental management functions of government in the Swedish Environmental Protection Agency (*Naturvårdsverket*). This has 500 employees and three main functions:

- Collecting and compiling data
- Developing environmental policy for the government
- Implementing the government's environmental policies

The Swedish EPA has an important coordinating role as host organisation for the Environmental Objectives Council, a committee representing the heads of a wide range of government agencies that sets objectives for Swedish environmental policy and develops implementation plans to achieve the targets.

The objectives are scoped at three levels of generality. The overall or 'Generational Objective' is 'To hand over to the next generation a society in which the major environmental problems in Sweden have been solved, without increasing environmental and health problems outside Sweden's borders'. The first half of this was passed through the parliament in the year 2000 via proposition 2000/01:130 – the qualifier about not exporting problems was added later. Supporting this overall objective, the government proclaimed 15 national environmental objectives which relate to environmental quality and ecosystem functions (today there are 16 – they are listed in Figure 8.5). The path to these environmental objectives is guided by the identification of a number of 'step goals' for each environmental objective, typically smaller in scope and easier for a national government to implement. For example, the national environmental objective of 'limited climate impact' is currently defined as the prevention of a 2°C temperature increase, and stabilisation of the atmospheric carbon dioxide concentration at 400 parts per million. Since this is an international

	Limited climate impact		Good-quality groundwater
	Clean air		A balanced marine environment, flourishing coastal areas and archipelagos
	Natural acidification only		Thriving wetlands
	A non-toxic environment		Sustainable forests
	A protective ozone layer		A diverse agricultural lanscape
	Safe radiation levels		Magnificent mountain environments
	No eutrophication		A good built environment
	Flourishing lakes and streams		A rich diversity of plant and animal life

Figure 8.5   Sweden's national environmental objectives (authors' translation – see www.miljomal.se)

matter, to make this objective clearer and practical in the national context, the step goal is that Sweden's net greenhouse gas emissions should be 40% lower in 2020 than they were in 1990.

The Environmental Objectives Council created a process for monitoring quantitative environmental quality indicators for each objective and goal, and has responsibilities to follow up these targets and to create intervention plans to ensure they are met. In many ways, this has created a kind of national environmental management system (EMS) akin to the requirements of ISO14001 (see Section 8.6), but in contrast to many corporate EMSs, the national goals are very ambitious and most are far from being reached, sometimes on account of factors outside Sweden's jurisdiction.

The Swedish EPA publishes an annual analysis of the trends associated with the quantitative indicators for each environmental objective. Every four years, a detailed analysis of the underlying causes of the trends is coordinated by the Swedish EPA in order to guide the government's intervention plans towards meeting the generational goal. The draft targets and plans proposed by the Environmental Objectives Council have to be presented to the government before they become policy, but they carry considerable weight as consensus statements of the organisations tasked with implementing such plans.

8.4 Statutory Environmental Impact Assessment

One of the important environmental management tools in use around the world is the environmental impact assessment (EIA) process. EIA is performed for a wide suite of developments ranging in scale from the renewal of suburban sewage pumping stations to the construction of international airports, and the scope of the planning work and documentation roughly corresponds to the scope of the development in question. The overall aim of EIA is typically to enable governments and the citizens they represent to determine whether an activity should be allowed in a certain place.

Unlike many of the other pieces of information which can influence environmental management in the government and corporate sectors, like LCA and ecological footprint analysis (see Chapter 7), EIA is strongly integrated in legislation, compelling organisations to gather and publish a set of information before a decision can be made regarding permission to interfere in the environment. Many countries have some form of EIA legislation at the national, state and local levels. Consequently, many thousands of environmental professionals are working in the EIA process around the world, including those who review them, those who write them and those who do the underlying environmental research. Many of these professionals are private sector employees.

The central document in an EIA is an environmental impact statement (EIS), though it can have other names (e.g. environmental impact report [EIR] under Californian law). This typically ranges in scale from the equivalent of a typical large-format university textbook, to a series of such books including various technical appendices – the scale depends on the complexity of the development being assessed and its potential to do harm. The overall process for writing an EIS is shown in Figure 8.6.

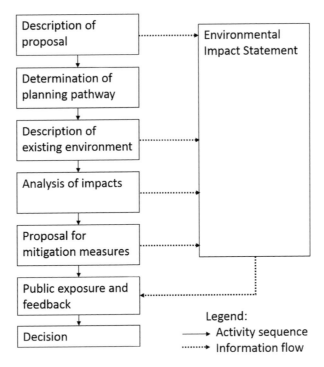

Figure 8.6   EIA process overview

The first step in an EIA is about clearly describing the development or action that is proposed. This means ensuring that the proponent of the action is clear about its scale. This might sound obvious, but it is important as insurance against the risk of 'scope creep' – an important project management hazard. (For example, it is not uncommon in infrastructure developments that the scope of a project will expand over time as stakeholders try to add elements to some major project that seems to be headed for approval.) Identifying the intended location of a development is critical at this stage since it determines which stakeholders are relevant to the proposal. This includes stakeholders in the fields of planning and policy such as decision-makers in government, stakeholders in the local community who may be directly affected by land use change or pollution and other stakeholders who may depend on shared resources in the region.

Identifying the location may also have important influence on the appropriate planning pathway – in other words, who gets to decide and what they get to see. For example, if the development is placed in the proximity of an endangered ecological community, it might mean that the proponent has to seek approval from a decision-maker higher up in the political hierarchy. The EIA would have to carefully describe the existing environment so the decision-maker can place the extent of the damage the proponent wants to cause in the context of the value of the existing environment. Obviously, constructing petrochemical infrastructure on degraded industrial land is of less concern than building it in a pristine environment. This is for example reflected in Swedish regulation SFS 2013:251, which says that if you should ever want to extract crude oil in certain high-altitude regions delineated in the

Table 8.3	Aspects of the Severity of an Environmental Impact (After Antweiler, 2014)	
Operation	Temporality	Magnitude
Incremental	Duration	Spread
Additive	Continuity	Intensity
Synergistic	Immediacy	On-site
Antagonistic	Frequency	Off-site
Probable	Regularity	
Uncertain	Reversibility	

Swedish Environmental Code, the action is classified 'A' – it has to be reviewed by a specialised court with the authority to consider matters of national importance. On the other hand, if you want to do it in other locations, it is only classified 'B', so you only have to take your EIS to the county officials. (This is a purely theoretical example, as Sweden does not currently produce crude oil.)

The other thing the analyst should be aware of at this point in an EIA is that in an attempt to focus regulatory resources on the issues that are considered the most important, many statutory planning processes have established a host of exemptions for particular kinds of actions. In the United States, there are long lists of 'Categorical Exclusions' to the NEPA Act, which mean, for example, that federal road work does not require an EIS if the work is to repair an existing road, or the federal government is providing less than $5 million in funding, or for one of many other reasons listed in Title 23 of the Code of Federal Regulations.

Depending on the activity that is to be assessed, many different kinds of impacts may be relevant. Obvious examples are pollution of the air, water and soil. Even at low concentrations – below any kind of toxic threshold – odorous gaseous emissions may be a concern. Local residents may be more concerned about other aspects that impact the amenity of the local environment, such as noise and vibration, light pollution and road traffic. Impacts on wildlife and endangered ecosystems may arise, for example via land clearing for the establishment of new facilities. Impacts of any kind should be considered and described in a variety of terms to identify their severity. Some of the less common among the terms suggested in Table 8.3 are discussed here. For example, an impact may occur as a single step, or the change may be *incremental* and continue indefinitely (compare erosion associated with construction versus erosion caused by continuous agricultural operations). The effects of multiple interventions may be *additive* (for example, multiple sources of the same contaminant), or may exceed the sum of their impacts (if one hectare of forest is removed in many places within an ecosystem, one may expect the fragmentation to cause a *synergistic* negative effect on threatened species that exceeds the effect of removing the same forest area all joined up). On the other hand, it may be possible for the total effect to be less than the sum of individual (*antagonistic*) effects under some other circumstances.

One of the more difficult aspects to be handled in an EIA is the matter of heritage protection. Old infrastructure may be practically useless and even dangerous from a technical point of view, but be considered a significant marker of the history of settlement in a location. If this is the case, it may be appropriate to put some of the infrastructure in a museum, or to make some kind of interpretative monument on site, to preserve something of its heritage value. Identifying the appropriate level of preservation may involve dialogue with and the need to get approval from heritage management authorities prior to the submission of the EIS to the ultimate decision-maker (for example, a national heritage management board may have to approve a heritage management plan in an appendix to an EIS, prior to submission of the EIS to a state government), with corresponding increased time demands on the EIA professional and the overall EIA process.

Mitigation measures such as these heritage management options may need to be proposed for all sorts of impacts identified in an EIA. Four general outcomes can be imagined. The impact:

- can be avoided via selecting an alternative technology that does not cause the problem in the first place, or an alternative location for where the activity does not matter;
- can be mitigated by some kind of on-site treatment or barrier;
- can be physically remedied, after it has occurred; or
- cannot be mitigated, but the victims can be compensated, for example financially or by granting them land elsewhere.

Providing a robust assessment of such diverse indicators of environmental impact as air quality and the survival of endangered species, and working out how these impacts can be mitigated, requires a wide range of talent. Therefore, EISs are typically produced by a team of professionals with diverse skills. For example, specialised skills may be needed to produce quantitative estimates of local residents' exposure to airborne contaminants near a development, using Gaussian plume models of the spread and dispersion of a gaseous emission diluted by local wind conditions and in the topography of the local environment. Such skills are very different from those held by heritage professionals, or professionals in communication whose input may be required as part of scoping a development, or getting community feedback on alternative options. EISs are typically produced under a contract provided by the proponent of the activity, to one of the large environmental management consulting firms, who are often supported by subcontracting to experts with more specialised skillsets.

Many criticisms of EIA are valid. The process has not had the degree of influence that was initially hoped for by the people who introduced it in Sweden and the United States. This is because the process is administered by existing power structures, the EIS is typically written by the (private or public) proponent of a development and public participation in EIA can be ineffective (Morgan, 2012). The focus on the proposed development by the proponent can mean that alternative developments are not properly explored, and that upstream and downstream issues that might be the normal concern of an LCA are given little consideration compared with local, on-site impacts. On the other hand, one of the key strengths of the EIA process is that it exposes decision-making to public scrutiny. In a country with freedom of speech and a flourishing media sector, this means that citizens can in principle gain awareness of planning activities, review the EIS and criticise it if it seems to endanger values that they hold.

8.5 Public Environmental Reporting

One of the ways in which governments and private companies are exposed to pressure to improve their environmental performance is via public environmental reporting. Public Environmental Reports (PER) go by various names and the kind of content can vary. Reviewing time series data for indicators in such reports provides useful feedback on the degree of success of environmental initiatives by governments and corporations. Governments at different levels (national, state, local) produce such reports, often called 'State of the Environment Reports' or similar names. Australia was a pioneer in this field when it published the 550-page report 'Australia: State of the Environment 1996'. The USEPA produced its first 'Draft Report on the Environment Technical Document' as a hard copy in 2003. This was revised in 2008 and called simply 'Report on the Environment', providing data on the USEPA's environmental quality monitoring work around the country. Since 2015, this has been replaced by an online report which is updated continuously (cfpub.epa .gov/roe). Currently, it includes data for 85 different environmental quality indicators, representing pollution concentrations in the air, water and soil; the condition of ecosystems and the biodiversity they support; and the exposure and health consequences of human exposure to contaminants.

In the private sphere, another common label for a PER is 'Corporate Responsibility Report' or 'Corporate Social Responsibility Report', though these sometimes ignore the natural environment. In some cases, the publication of an annual PER is required by government regulations. For example, in the European Union, all publicly listed companies with more than 500 employees are required to publish an annual report on their policies, risks and results in relation to social and environmental impacts, human rights, diversity and corruption. When the legislation forcing this change on the corporate sector was passed in 2014 (EU Directive 2014/95/EU), 2,500 European companies were already delivering such reports or combining them with their annual financial report. It was estimated that a further 4,500 companies would have to do this in 2017 when the legislation took effect (Fried, 2014). In other cases, organisations voluntarily choose to report on their environmental performance in order to establish their credibility with their peers and the citizens in the places in which they operate. Another driver is that potential investors may view such reports as necessary to understand whether a company is exposed to reputational risks. A survey by an accounting firm showed that about 90–95% of the largest 250 corporations on the planet are publishing PERs, usually as part of their mainstream annual report (Blasco et al., 2017).

When different private corporations publish environmental reports, the kinds of indicators that are relevant may vary considerably depending on the scale of the business, the products or services it sells and the locations in which it operates. A non-profit organisation, the Global Reporting Initiative (GRI) (www.globalreporting.org), has provided important support for corporations looking for a flexible standard on how to produce PERs, and the EU Directive supports the use of the GRI's guidance documents.

8.6 Corporate Self-Regulation: Environmental Management Systems

Imagine yourself as the manager of an oil refining company, worried about what you can do to prevent environmental damage by your company. It is very difficult to visit all your production sites around the world regularly – the company is too big for that to work, and anyway you are not a qualified environmental engineer. So you have to create a system of management which lets people know what you want them to do about avoiding pollution, and which tells you whether they are doing it. This is the idea behind environmental management systems (EMSs). In and of themselves, they do not create environmental progress, although when they are coupled to bold goals and endorsed by executives, they can be of great assistance. What they do primarily is to establish a feedback loop linking performance monitoring and planning within a company.

8.6.1 The Plan-Do-Check-Act Cycle

The Plan-Do-Check-Act (PDCA) cycle was popularised by the work of American engineer, management consultant and academic W. Edwards Deming, although he always credited another fellow engineer, Walter Shewhart, with the idea. It is summarised in Figure 8.7. The basic elements are: planning – determining performance criteria and the means to achieve them; doing – implementing the plan and gathering data on the results as you go; checking – analysing the data from the previous step and looking for deviations from the intended results; and acting – deciding whether the goals, performance criteria or other aspects of the process need to be adjusted going into the next iteration of the planning step.

Shewhart saw this cycle as a practical approach to inductive reasoning – a basic scientific method. Experimental science involves creating a hypothesis, performing an experiment to test it and evaluating the results. These parallel the plan, do and check parts of the Shewhart cycle, respectively. Deming promoted the Shewhart cycle to his many followers in Japan during its reconstruction after the Second World War as a way to ensure the quality of industrial production, and he is credited as having been the most influential non-Japanese person in the 'Japanese economic miracle' in which Japan became a major technology exporter. The value of the Shewhart cycle as a tool in quality management is manifested in its incorporation into quality management standards like ISO9001 (ISO, 2015a).

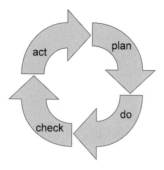

Figure 8.7   The Shewhart or Deming cycle

While the Shewhart or PDCA cycle was originally promoted as a way to improve the quality of industrial products, it can in principle be used to ensure that an organisation is moving towards many other kinds of objectives, including environmental objectives. This is why it is also a key element of ISO14001, the global standard for environmental management systems (ISO, 2015b). An environmental management system (EMS) is 'a formal system and database which integrates procedures and processes for training of personnel, monitoring, summarizing, and reporting of specialized environmental performance information to internal and external stakeholders of a firm' (Sroufe, 2003, p. 426). Elements of each of the four components of the PDCA cycle are shown in Table 8.4, tailored for a corporate EMS.

8.6.2 Strengths and Weaknesses of Environmental Management Systems

The genius of an EMS is that it potentially creates a virtuous 'learning circle', causing improved performance. Establishing an EMS is a standard way for executive management to exert influence on corporate environmental performance. Third-party audits of EMS systems can be employed to ensure compliance with ISO14001, meaning that an EMS is performing its role in ensuring staff and management are acting in compliance with internal and external environmental performance goals. Unfortunately, it is also possible to observe EMS being used as proof that an organisation or even a product is environmentally sustainable. Of course it does nothing of the sort, since the extent to which it influences the environmental performance of an organisation is dependent on the level of ambition executive management expresses in the 'act' step of the Shewhart cycle, when it reviews and determines the environmental goals for the organisation. The existence of an EMS in an organisation does not per se guarantee above-average environmental performance. There are examples of companies which have third-party certified EMSs, but which use the EMS merely as a way of avoiding environmental risk, rather than setting ambitious goals to improve their performance beyond legal compliance.

Table 8.4 Key Elements of an Environmental Management System (EMS)	
PDCA step	EMS activity
Plan	• Secure written management commitment to the EMS. • Define environmental goals and targets for the organisation. • Identify human and budgetary resources to run the EMS. • Have a kick-off meeting for the EMS team. • Initiate communication with stakeholders and define the views of interested parties. • Identify relevant legal requirements. • Identify relevant environmental issues. • Describe relevant company operations. • Develop environmental management plans (EMPs). • Establish a monitoring programme. • Collate an EMS manual.
Do	• Educate staff regarding their environmental responsibilities. • Operate and maintain environmental protection equipment. • Collect operational data relating to environmental performance.
Check	• Assess the operational data – e.g. how well did the company perform? • Audit the operation of the EMS itself – e.g. did all the information come in?
Act	• Executive management reviews the outcomes of 'Check'. • If performance is satisfactory, goals can be made more ambitious. • If performance is unsatisfactory, planners can be directed to improve plans and ensure targets are met.

8.7 Review Questions

1. What are the most important international environmental laws on toxic chemicals?
2. How does international environmental law come into force? Consider also the operation of national governments in the United States and Sweden.
3. You want to import two tonnes/year of a new chemical into France. What happens next under REACH?
4. What is a Deming cycle and how is it used in environmental management?
5. What are the elements of a statutory environmental impact assessment?
6. Explain the differences in legal basis and scope between a life cycle assessment (Chapter 7) and an environmental impact statement.
7. If Sweden's environmental objectives are considered part of a national EMS, what are the other elements of the system?

9   Decision-Making

9.1 Introduction

The examination of how people make decisions encompasses a wide range of business management, operational, political and psychological research. Rather than reviewing the whole literature, this chapter focusses on providing practical tools for technical professionals, and on some of the cultural and ethical problems that arise in professional life.

Engineers and scientists have decisions to make about how to pursue particular projects and how to prioritise alternative options. Sometimes they are employed as consultants to institutions needing external, impartial support for decision-making processes. 'Multicriteria analysis' (MCA) and 'multicriteria decision aiding' (MCDA) are established tools which analysts frequently use to help people to make decisions. These tools are applied to the practical challenges of decision-making in fields like infrastructure development and product eco-design. They generally provide a logical framework to manage trade-offs and to combine multiple perspectives in the presence of uncertainty. Stakeholders may have different perspectives and priorities regarding factors such as the social, financial and environmental performance of alternative options, and evaluating each of these factors can make the decision an interdisciplinary problem. Uncertainty can arise, not only in terms of whether stakeholders who are relevant to the decision are adequately represented, but also in relation to the performance of different, alternative options. In the face of uncertainty and conflicting perspectives, performing an MCA can be as much a learning experience as a decision process for the people engaged in it.

Liberal democracies and private companies may sometimes have diametrically opposed expectations about how open decision-makers and their analysts should be when making a decision. Nevertheless, both can benefit from having a clear way of structuring a decision-making process, and a record explaining how a decision was made. Even if the record is confidential, it can be helpful to a decision-maker faced with many decisions each year to have a record of what they were thinking, if decisions need to be subsequently reviewed. This chapter is intended to provide a basis for structuring decision-making processes and explaining decisions. Note that the words 'alternatives' and 'options' are used interchangeably here.

9.2 Types of Decisions

Many typologies can be used to describe the aims of different decisions. It has been suggested (Rowley et al., 2012) that the following set of problem formulations covers the decisions that analysts frequently handle:

1. Sorting problem e.g. 'Which category does this alternative belong in?'
2. Choice problem e.g. 'Which of the alternatives available to me is most suitable?'
3. Ranking problem e.g. 'In what order should I consider these alternatives to be?'
4. Description problem e.g. 'What are the consequences of choosing this alternative?'

In each of these situations, it may be that several alternatives are of equivalent value, so it is not the case that (2) or (3) must always pick a unique 'winner'. Problem (4) is sometimes seen as a subset of one of (1) to (3) (Roy, 1981). The problem formulation is relevant to the selection of decision-support method, so the analyst and the decision-maker should discuss this early and record their thinking.

9.3 Defining the Decision-Maker

The easiest situation to handle is one in which the analyst (for example, a recently graduated engineer) can identify a single decision-maker (for example, their boss). In other words, there is one person whose values have to be incorporated in the decision process. This person may be the same as the analyst, but in the case of consultants providing decision support for industrial or government clients, the analyst is frequently required to be completely impartial.

However, in many situations, the decision process is intended to reflect the priorities of a group of people affected by the decision, often called 'stakeholders'. This creates the challenge of ensuring that the committee or panel of people entrusted with making a decision is in fact the appropriate group or that it reflects all relevant perspectives. It also presents the problem of how to encompass the breadth of opinion which may be present within such a group. For example, a simple arithmetic average of the group's opinion regarding the weighting of some criterion may be easy to calculate, but if there are widely diverging opinions within the group, such an approach may satisfy nobody within it, and some kind of sensitivity analysis will be required. The principle of 'one vote, one value' is enshrined in democratic thought, but it may be inappropriate in situations where a committee for arbitrary reasons is stacked with people who represent one point of view. For example, if a business management committee has several absentees from one area of the business, due to ill health or parental leave, it may be necessary to compensate for imbalance in the committee, for example by mathematically adjusting the voting power of those present, or excluding some members of the committee from voting. In any case, the analyst will have to be open about the options and the procedure for a decision process to be successful, and the approach should be justified in the written record of the decision process.

9.4 Identifying Alternatives

Lundie et al. (2008) described a number of methods to ensure analysts are not stuck within their own field of view when a group of stakeholders wishes to identify alternative policies or engineering solutions for assessment:

- *Brainstorming*: The group is encouraged to express and record random ideas. Later, the associated pros and cons are discussed.
- *Brainstorming with SWOT*: The group considers the strengths, weaknesses, opportunities and threats (SWOT) of the status quo as a starting point for identifying alternatives, or the SWOT can be the basis of a discussion which provides an initial screening of ideas derived from an initial brainstorm. Strengths and weaknesses relate to the intrinsic characteristics of the status quo or the alternative, while opportunities and threats are external.
- *Negative brainstorming*: The group imagines the worst possible outcome, then works methodically back to how it can be prevented, taking note of the alternative policies or solutions that prevention will require.
- *Lateral thinking*: The group uses experience from other fields to create solutions. For example, small-scale drinking water supply systems have been built using experience in air-conditioning system design, after observing that atmospheric water vapour drips from them under humid conditions.
- *Systematic inventive thinking*: This is based on the idea that there are common elements to many creative processes, and that group brainstorming may be too random to be effective (Goldenberg et al., 1999). Thinkers are encouraged to invert principles they might expect to use for innovation, starting with existing alternatives and modifying them by, for example, 'thinking inside the box' and imagining 'function follows form', rather than the opposite.
- *Pragmatic design*: Like systematic inventive thinking, this means looking for new ideas based on previous experience, current ideas and possible innovation on them as ways to identify solutions (Mitchell & White, 2003).
- *Backcasting*: This is quite different to the previous methods in this list. Starting from criteria that describe a desired scenario at a time well in the future, backcasting requires the user to think 'back' to an earlier time in the future, to think about the prerequisites needed at that point to reach the scenario. The process iterates back to the present day to suggest the potential alternatives to business as usual which are needed now.

9.5 Selecting Criteria

In decisions concerned with the sustainability of alternative options, contributions to climate change (usually aggregated using an LCA method) are frequently chosen as a proxy indicator for environmental damage, but many other LCA indicators and other environmental indicators may be relevant. Likewise, indicators of financial burdens (e.g. capital cost, net present value), social impacts (e.g. net jobs created) and other issues may be relevant. The analyst may have useful ideas about what matters but at the end of the day, the decision-maker needs to agree with the selection of criteria in a decision. The larger the number of criteria to be evaluated in an MCA process, the larger the workload for the analyst. On the other hand, the number of criteria must be adequate to cover the key issues that are relevant to the stakeholders. The criteria should be:

1. exhaustive – they should not miss any important characteristics of the alternatives
2. minimal – the number of criteria must be manageable for the human beings using them
3. monotonic – to allow consistent comparison of partial and global preferences, marginally improving performance of an option should always marginally improve the criterion
4. independent – criteria should represent different issues and avoid double counting

Naturally, the decision-makers need to be in agreement about the criteria with the analyst, so the criteria must be brought up for discussion early in an MCA process before effort is wasted evaluating criteria that stakeholders do not consider relevant.

9.6 Evaluating Criteria

This is the part of the decision process which may rely heavily on analysts with the technical knowledge to estimate the performance of potential alternatives, but there may also be criteria that require subjective ratings or other inputs from stakeholders. Depending on the number of alternatives and criteria, it may be wise for the analyst to include a preliminary screening of alternatives. If there are many alternatives, it may be possible to eliminate several by, for example, applying veto (or 'dominance') thresholds to the criteria. Consider the supply of water to a growing coastal city in a country dependent on coal for electricity. Perhaps the options for expanding the supply include recycling sewage, desalinating seawater and dehumidifying the air. If greenhouse gas emissions are a criterion for consideration and renewable energy is unavailable, the analyst could consider that while the emissions from recycling and desalination are the same order of magnitude, the emissions from dehumidification are three orders of magnitude worse (Peters et al., 2013). If dehumidification performs so poorly against this one criterion, it may be eliminated from contention without the effort of evaluating it against the other criteria. So, in this example, a preliminary screening allows the number of options to be reduced by a third.

9.7 Aggregating Preferences

There are many ways to aggregate preferences. Ishizaka and Nemery (2013) provide an introduction to this field. MCA methods are generally divided into two main classes: approaches based on a synthesising criterion (calculating some kind of total score for each alternative and then comparing scores) or on a synthesising preference relational system (for example performing lots of pairwise comparisons and aggregating these comparisons). To make sense of what this means, this book describes four of the most popular approaches.

9.7.1 Synthesising Criterion

9.7.1.1 MAUT

There is a broad class of methods that can be grouped under the heading of 'multi-attribute utility theory' (MAUT) and that uses weights for aggregating preferences. One of the simplest approaches is the weighted sum. It can easily be used for the second and third types of problems listed in Section 9.2. It is embodied in the formula:

$$U_j = \sum_{i=1}^{m} w_i a_{ij} \qquad j = 1, 2, 3 \ldots n \tag{9.1}$$

where U_j is the utility (benefit) of alternative j (one of n alternatives), w_i is the weighting associated with criterion i and a_{ij} is the utility of alternative j against criterion i. Generally, the weights should add up to one. Comparing two options, we have:

$$\text{'}j_1 \text{ is preferred to } j_2\text{', i.e.: } j_1 \mathbf{P} j_2 \Leftrightarrow U_{j1} > U_{j2}$$

or

$$\text{'}j_1 \text{ and } j_2 \text{ are indifferent', i.e.: } j_1 \mathbf{I} j_2 \Leftrightarrow U_{j1} = U_{j2}$$

This formula is simple per se. The challenges in using MAUT arise when the analyst has to place the performance of the alternatives on a common scale (a_{ij}) and to acquire weights (w_i) from decision-makers and stakeholders.

A key issue here concerns the way in which weights are obtained. Many analysts incorrectly ask decision-makers to assign weights without indicating to which scale the weights refer. So, for a simple example, if the price of a house is to be considered along with its size, the size could be expressed in square feet or alternatively in square metres. Assume that in either case the price is in dollars. If the analyst suggests the buyer just gives X% of the weight to price and 100−X% of the weight to size (i.e. a dimensionless importance coefficient for w_i in Equation 9.1), the importance of the area to total U will be about nine times higher if the size ($a_{size,i}$) is expressed in square feet (rather than square metres). The analyst should really ask what a square foot is worth in dollars to the buyer. (Or, in a country with the metric system, what the price of a square metre should be.)

Another way around the weighting problem in this example is to use scaling to make data more comparable. The four simplest ways to do it are:

1. 'Zero-max': scores (raw a_{ij} values) are scaled according to their position between zero and the best score for a criterion.

2. 'Min-max': scores are scaled to fit proportionally between the lowest and highest score for a criterion.
3. Scores are expressed as a fraction of the sum of all scores for a criterion.
4. Scores are expressed as a fraction of the square root of the sum of the squares of scores for that criterion.

There is no universal preference for one of these, but the second option suffers a loss of proportionality and is very sensitive to the selection of alternatives. If an unscrupulous decision-maker introduces an unrealistically poor alternative (or 'straw man') into the decision, which performs terribly against one of the criteria, this will have the effect of 'bunching' the other scores for this criterion at the high end of the scale, reducing the discerning power of the criterion compared with the others.

In LCA, a version of the first option is often employed and called 'normalisation', in which an amount of emissions is compared with a local, regional or global total amount of emissions. This creates more difficult problems for the analyst if there are only regional estimates for one emission and global emission estimates for others. A similar problem emerges if some impact only occurs in the local environment – the emissions causing that impact should arguably be normalised against the total local emissions, while the other emissions cause global impacts and need a global total for normalisation. Weighting these normalised criteria then requires the decision-maker to weigh local impacts against global ones, a difficult feat which is not adequately presented to decision-makers by just asking them to assign a percentage to each.

We may have to deal with both positive and negative indicators (in the latter, higher scores are worse – greenhouse gas emissions is an example) in a single MCA. There are a couple of possible ways to reverse the sense of the indicator, but there is a consensus that for methods like MAUT, inverting the scores is superior (Rowley et al., 2012; Zarghami & Szidarovszky, 2011), since it preserves proportionality, so for zero-max scaling:

$$a'_{ij} = \frac{\min(a_{ij})}{a_{ij}} \tag{9.2}$$

Note also that the utility a_{ij} does not have to be a linear relation of some kind of performance data, so, for example, one could imagine that someone wishing to buy a house might be a little flexible on the price but sees an asymptotic relationship between the price of a house and its value to her decision-making (see Figure 9.1).

Constructing such non-linear utility functions as this may be difficult in many practical corporate and governmental situations, as it has to represent the utility of performance against a criterion for all the relevant decision-makers.

9.7.1.2 AHP

The 'analytical hierarchy process' (AHP) was developed in the United States and is very popular, probably because it seems easy for stakeholders to understand the weighting process using simple verbal phrases. On the other hand, at least in its more popular forms, its mathematical basis is somewhat arbitrary. AHP begins like many MCA methods by defining the aim of a decision process, the alternatives to be

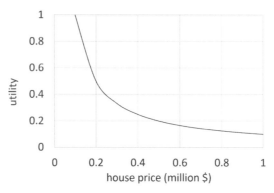

Figure 9.1   Negative marginal utility function

considered and the criteria for evaluating them. Subsequently, the relative importance of the criteria is determined by a particular weighting procedure. Then the performance of the alternatives can be described and a single score calculated for each.

The weighting step in AHP creates a square 'comparison matrix' for the criteria. Let us stay with the house-buying example. We use a standard nine-level preference scale as shown in Table 9.1 (only five of the levels are shown; the other, even, values fit between the expressions shown). We set up a verbal preference scale as shown. Imagine that the buyer has three criteria, and rates them as shown in Table 9.2.

We read the data in Table 9.2 from the left to the top, i.e. the buyer stated that location is very strongly more important than the size of the house. In general, to generate this table, the analyst has to ask the decision-maker $(n^2-n)/2$ questions, where n is the number of criteria. This is because it is assumed that the preferences

Table 9.1 AHP Fundamental Preference Scale

Preference judgement	Preference value
Both criteria are equally important.	1
This criterion is moderately more important than the other.	3
This criterion is strongly more important than the other.	5
This criterion is very strongly more important than the other.	7
This criterion is extremely more important than the other.	9

Table 9.2 Example AHP Comparison Matrix

	Size	Price	Location
Size	1	1/3	1/7
Price	3	1	1/5
Location	7	5	1

Table 9.3 Criteria Weights

	Geometric mean	Normalised mean
Size	1.139	0.223
Price	1.613	0.316
Location	2.351	0.461
Sum	5.103	1.000

Table 9.4 Alternatives Compared on Size

	House A	House B	House C	Normalised mean
House A	1	0.5	1	0.25
House B	2	1	2	0.5
House C	1	0.5	1	0.25

Table 9.5 Outcomes of Simple AHP Example

	Comparison matrix			Weighted matrix		
	House A	House B	House C	House A	House B	House C
Size	0.25	0.50	0.25	0.056	0.112	0.056
Price	0.550	0.240	0.210	0.174	0.076	0.066
Location	0.135	0.367	0.498	0.062	0.169	0.229
Sum:				0.292	0.357	0.352

also hold when inverted – in the example: $n = 3$ and the size of the house is very strongly less important than the location.

The matrix in Table 9.2 has to be turned into a single weight for each criterion. An easy way to do this involving the geometric mean is described here (based on Munier, 2011). First, the geometric mean of each row is computed and then the values are normalised by their sum, so that they add up to one, as shown in Table 9.3.

The same procedure of taking the geometric mean and normalising is applied to the comparison of alternatives for each criterion. So, for example, the decision-maker is asked, 'considering the size of the alternatives, how strong is your preference for house A over house B?' This creates a series of matrices like Table 9.4 which concerns the size of the different houses. One table like this is made for each criterion.

With data for the other two criteria, the analyst can weight the performances of the alternatives using the weights in Table 9.3, as shown in Table 9.5.

In this case, House B is preferred by a small margin, so it would be worthwhile to test the robustness of the preference by sensitivity analysis. Software packages exist to simplify the use of AHP and make such sensitivity testing easier. They can also employ variations on the method such as different ways of deriving the priorities or checking for errors. Traditionally, AHP uses the computation of eigenvectors of matrices as part of assessing whether the matrices are consistent. Consistency means, for example, that if a buyer considering one criterion likes House B twice as much as House A, and likes House A three times as much as House C, then she must like House B six times as much as House C. If the matrices are inconsistent, the analyst needs to go back to the evaluation process and try to ensure they are consistent.

9.7.2 Synthesising Preference Relations

Not everybody is happy with the idea of modelling preferences using a value function or a utility function (e.g. Roy, 1990). A major issue is the question of whether a criterion should be allowed to compensate for another criterion (which is discussed further in Section 9.7.5). There may be uncertainties in the mind of the decision-maker which are hard to model with a synthesising criterion. For example, it may be difficult to turn ordinal information (e.g. X is bigger than Y) into cardinal data (e.g. $X = 10$, $Y = 3$) for an indicator. Or it may be difficult to normalise indicators when they operate on different geographical scales (e.g. local versus global). For reasons like these, various MCA methods built on synthesising preference relations (pairwise comparison) have been constructed. We now turn to two of the most popular ones: PROMETHEE and ELECTRE.

9.7.2.1 PROMETHEE

PROMETHEE is an acronym for *Preference Ranking Organisation METHod for Enriched Evaluation*. This pairwise comparison method was developed in Europe and has been in practical application since the early 1980s. For detailed information about this method, the reader is referred to Brans and Mareschal (2005). Briefly, the idea is to compare each pair of alternatives on the basis of each criterion, and to identify the degree of preference for the preferred alternative. Preference 'flows' are then added up.

The difference in performance of alternatives is placed on a 0–1 scale using one of several preference functions that are chosen to reflect the information available to the decision-maker. These are summarised in Figure 9.2. In each of the six cases, the x-axis shows an increasing difference between alternatives. The y-axis shows the value of the preference for the better alternative.

The simplest 'usual' function – type 1 – is sometimes called 'strict preference', and it has the drawback that even tiny differences between alternatives result in complete preference for the better alternative. For example, you believe price and size are equally important among a group of houses, and if a house that is only 100 m^2 costs $999,999, while another that is 200 m^2 costs $1,000,000, this function would give all the weight of your financial vote to the cheaper house, and all the size vote to the bigger house, resulting in a tied vote between these two. On the other hand, using the type 6 'Gaussian' preference function would recognise that the prices are almost the same but the sizes are very different, and the bigger house would win the comparison between the two. Using the Gaussian approach means that the decision-maker needs to supply a value for *s*, the midpoint on the Gaussian curve:

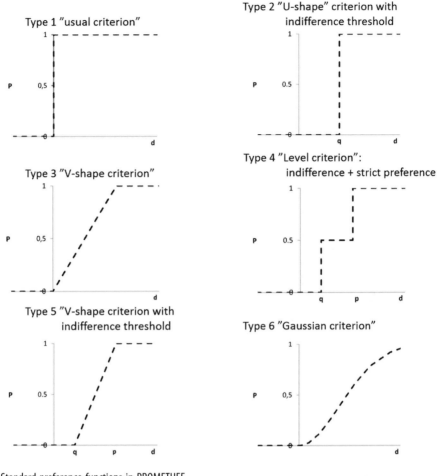

Figure 9.2 Standard preference functions in PROMETHEE

$$\text{Preference} = 1 - e^{-\frac{d^2}{2s^2}} \tag{9.3}$$

where d is the difference between the performance of two alternatives. Some of the other standard preference functions require the provision of an indifference threshold q on the d-axis (types 2, 4 and 5) or a threshold p for full preference (types 3, 4 and 5), so identifying these will mean extra work for the analyst and the decision-maker. On the other hand, making the preference function fuzzy (unlike types 1 and 2, capable of having a value other than 0 or 1) makes it more likely that the analyst will avoid the arbitrary impacts of marginal changes in the data used in the MCA.

Here is a mathematical example of this effect using PROMETHEE. Suppose you are considering three houses on the market and what matters most to you are their price, size and location. You have rated the location of each house on a five-point scale and consider each criterion about equal weight, as shown in Table 9.6.

Table 9.6	Example PROMETHEE Input Data		
Criterion	Price (thousands of $)	Size (m^2)	Location (five-point rating)
Weight	0.33	0.33	0.34
House A	998	100	4
House B	1,000	200	3.5
House C	1,500	150	3

Table 9.7	Single Criterion Preference Function Results					
	Type 1 criterion			Type 3 criterion		
	Price	Size	Location	Price	Size	Location
P(A,B)	1	0	1	0.02	0	1
P(A,C)	1	0	1	1	0	1
P(B,A)	0	1	0	0	1	0
P(B,C)	1	1	1	1	1	1
P(C,A)	0	1	0	0	1	0
P(C,B)	0	0	0	0	0	0

A preference table $P_j(d_j(a,b))$ is calculated using a preference function to present the differences between alternatives into a preference scale (Table 9.7). Here $d_j(a,b) = f_j(a)-f_j(b)$, which is calculated for the performance f of each positive alternative with respect to criterion j, but for the price (a negative indicator) instead $d_j(a,b) = f_j(b)-f_j(a)$. Note for the sake of illustration, P_j has been determined using both a type 1 criterion and alternatively a type 3 criterion. In the latter case, it was assumed that the preference thresholds (the difference between scores at which P goes to 1) were $100,000, an area of 30 m^2 and 0.5 on the location quality scale.

Multiplying the weights for each criterion by the preferences, we can calculate an aggregated preference (or 'multicriteria preference degree' = π) for each comparison, using formula 9.2, as shown in Table 9.8:

$$\pi(a, b) = \sum_{j=1}^{k} P_j(a, b)w_j \qquad (9.4)$$

Table 9.8	Aggregated Preference Function Results					
	Type 1 criterion			Type 3 criterion		
	... over A	... over B	... over C	... over A	... over B	... over C
Preference for A ...	-	0.670	0.330	-	0.347	0.670
Preference for B ...	0.330	-	1.0	0.330	-	1.0
Preference for C ...	0.330	0.0	-	0.330	0.0	-

where there are a total of k criteria denoted by j.

Then for each of the n (in this case, three) alternatives, we add the rows in Table 9.8 and subtract the columns. In PROMETHEE terms, this means subtracting the total negative preference flow from the total positive preference flow to calculate a net preference flow Ø for each alternative, as described by Equation 9.5.

$$\emptyset(a) = \frac{1}{n-1} \sum_{j=1}^{k} \sum_{\forall x} [P_j(a,x) - P_j(x,a)]w_j \tag{9.5}$$

Here 'a' is an alternative being compared in the MCA with each of the n alternatives (denoted by 'x' in the equation). The results of this example calculation are shown in Table 9.9.

This example illustrates the influence of small differences on preference relationships in PROMETHEE. Under the type 1 preference function, the positive preference flow to House A is larger than under a type 3 preference function, with the consequence that it is the highest-ranked alternative. From the other perspective, using a type 3 criterion, the effect of small differences is reduced, the positive preference flow to House A is lower, and this alternative is therefore ranked in second place.

9.7.2.2 ELECTRE

The ELECTRE method was originally developed by Bernard Roy (1968) of Paris Dauphine University, hence the French name: 'ELimination Et Choix Traduisant la Realité' (ELimination and Choice Expressing Reality). Like the other major methods described in this chapter, it has been the subject of much research and practical use, resulting in several variants for different kinds of problems:

ELECTRE I : This works for the sorting problem (see Section 9.2), finding a partial ranking of alternatives and a satisfactory subset of them by computing the sum of the weights for criteria for which alternative A outranks alternative B ('the concordance index') and a score for when A does not ('the discordance index'). These terms are described in more detail shortly.

ELECTRE II : This method introduces a veto threshold and orders alternatives (which is the ranking problem in Section 9.2).

	Table 9.9 Outcome of PROMETHEE Rankings							
	Type 1 criterion				Type 3 criterion			
House	Positive flow	Negative flow	Net flow	Rank	Positive flow	Negative flow	Net flow	Rank
A	0.67	0.33	0.34	1	0.5083	0.33	0.1783	2
B	0.665	0.335	0.33	2	0.665	0.1733	0.4917	1
C	0.165	0.835	−0.67	3	0.165	0.835	−0.67	3

ELECTRE III : To ELECTRE II, this adds preference and indifference thresholds, to eliminate the influence of small differences.

ELECTRE IV : This method is a development of III that can be operated without weighting the criteria.

A further suite of versions has been proposed, called ELECTRE TRI.

One of the main things all these methods have in common is that they avoid compensation between criteria (see Section 9.7.5). The rest of this section describes the most popular (Govindan & Brandt Jepsen, 2015) of these methods: ELECTRE III.

To evaluate whether option 'a' is preferable to option 'b' (*aPb*), we imagine four possible situations in which we consider whether *a* is at least as good as *b* (denoted *aSb*):

- *aSb* and not *bSa* (*a* is strictly preferred to *b*)
- *bSa* and not *aSb* (*b* is strictly preferred to *a*)
- *aSb* and *bSa* (*a* is indifferent to *b*)
- not *aSb* and not *bSa* (*a* is incomparable to *b*)

For *aSb* to be true, a sufficient majority of criteria should support the assertion that *a* is at least as good as *b* (the 'concordance principle') and none of the criteria should strongly oppose it (the 'discordance principle'). In ELECTRE III, the degree of concordance *(c)* and discordance *(d)* are calculated (between 0 and 1) by considering thresholds for preference *(p)*, indifference *(q)* and veto *(v)*, as shown in the example in Figure 9.3.

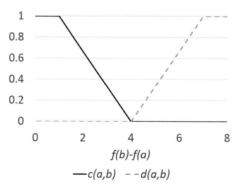

$f(b)-f(a)$

—$c(a,b)$ - -$d(a,b)$

Figure 9.3   Example fuzzy relationship – concordance and discordance for p = 4, q = 1 and v = 7

The simplest algebra for this relationship is perhaps provided by Li and Wang (2007). It is that for option *a* and option *b*, the concordance *c(a,b)*:

$$c_i(a,b) = 1 \text{ for } f_i(b) - f_i(a) \leq q_i \tag{9.6}$$

$$c_i(a,b) = 0 \text{ for } f_i(b) - f_i(a) \geq p_i \tag{9.7}$$

$$c_i(a,b) = \frac{p_i + f_i(a) - f_i(b)}{p_i - q_i} \text{ otherwise.} \tag{9.8}$$

This value denotes the extent to which for criterion *i*, it is true that *a* is at least as good as *b*. In the same vein, the discordance d(a,b) is defined as:

$$d_i(a,b) = 1 \text{ for } f_i(b) - f_i(a) \geq v_i \tag{9.9}$$

$$d_i(a,b) = 0 \text{ for } f_i(b) - f_i(a) \leq p_i \tag{9.10}$$

$$d_i(a,b) = \frac{f_i(b) - f_i(a) - p_i}{v_i - p_i} \text{ otherwise.} \tag{9.11}$$

This is the extent to which criterion i does *not* support the notion that a is at least as good as b. If the criterion i is negative (a higher value is worse), then we use the difference between the negative values, i.e. $(f_i(a) - f_i(b))$ in these expressions. A square matrix with n^2 entries (where n is the number of options) called the global concordance degree, $C(a,b)$, is defined as:

$$C(a,b) = \frac{1}{\sum_m w_i} \sum_{i=0}^{m} w_i c_i(a,b) \tag{9.12}$$

Where w_i is the weight given to criterion i and there are m criteria. The outranking matrix is then constructed with $S(a,b)$ equal to:

$$S(a,b) = C(a,b) \prod_{i=1}^{m} T_i(a,b) \quad \text{where} \tag{9.13}$$

$$T_i(a,b) = \left[\frac{1 - d_i(a,b)}{1 - C(a,b)} \right] \text{ if and only if } d_i(a,b) > C(a,b) \text{ and} \tag{9.14}$$

$$T_i(a,b) = 1 \text{ otherwise.} \tag{9.15}$$

Let us reimagine our house-buying example, keeping the criteria weights in Table 9.3. The initial data for this problem are shown in Table 9.10.

First, we calculate the differences between the performances of the options, reversing the sense of the comparison of price, since that is a negative criterion. This generates Table 9.11.

We now introduce thresholds for preference, indifference and veto (Table 9.12). These were defined mathematically earlier – in this example, the preference threshold for the size criterion is the point at which the house buyer feels there is a clear difference between the size of two houses, whereas below the indifference threshold, the buyer

Table 9.10 Decision Matrix			
Criterion	Size (m²)	Price ($million)	Location (quality)
Criteria weights	0.22	0.32	0.46
Action	Maximise	Minimise	Maximise
House A	90	0.8	1
House B	110	1.5	1.3
House C	100	0.9	0.9

Table 9.11 Normalised Matrix			
Criterion	Size (m^2)	Price ($M)	Location (quality)
Criteria weights	0.22	0.32	0.46
Action	Max	Min	Max
House A versus B	20	−0.7	0.3
House A versus C	10	−0.1	−0.1
House B versus C	−10	0.6	−0.4
House B versus A	−20	0.7	−0.3
House C versus A	−10	0.1	0.1
House C versus B	10	−0.6	0.4

Table 9.12 Example Thresholds			
Criterion	Size (m^2)	Price ($M)	Location (quality)
Preference (p)	10	0.1	0.5
Indifference (q)	5	0.05	0.2
Veto (v)	15	0.8	0.7

Table 9.13 Concordance Matrix			
	House A	House B	House C
House A	1	0.63	0.78
House B	0.68	1	0.68
House C	0.68	0.47	1

just does not care about the difference. If the size difference exceeds the veto threshold, then as far as the buyer is concerned, the smaller house is just not a contender. Using these thresholds enables us to generate a concordance matrix (Table 9.13).

In the case of the size criterion, we compare A with B and can see that the criterion should be maximised. In this case, A does not satisfy the criterion (the difference between A and B is larger than p), so we give it a score of zero (Equation 9.7). On the other hand, the price criterion is to be minimised, and A does satisfy this in relation to B. So, in this case, we give the full weight of the criterion to A (i.e. 0.32) since the difference between A and B is less than q (the difference is a negative number). Finally the location criterion is partially satisfied in this comparison, being less than p but more than q ($C(A,B) = 0.67$), so it gets two-thirds of the weight for size comparison (0.31 via Equation 9.8). The total of these scores for the comparison of A with B is then 0.63. In this way, Table 9.13 is generated.

In a similar manner, using formulas 9.9–9.11 and the data in Table 9.11 and Table 9.12, a partial discordance matrix (Table 9.14) can be generated:

The outranking (or 'preference') matrix is shown in Table 9.15.

Notice how there is a zero in the House B column for row House A. This is a consequence of the discordance matrix having (at least) one value of one for the performance of A compared with B, which is greater than the corresponding concordance index value of 0.63. So, the veto threshold of 15 m^2 has played the role of eliminating this score here.

Various means have been proposed for distilling a ranking from the preference table, taking into account some kind of threshold of difference. The basic approach is to add the preferences for an option (the total of the rows in the preference matrix) and subtract the preferences against it (the total for the columns). In this case we have:

A: $0.78 - 0.98 = -0.20$
B: $0.91 - 0.47 = 0.44$
C: $1.15 - 1.39 = -0.23$

Thus, House B is the highest ranked option and House C the lowest. Additionally, one can 'defuzzify' the preferences by

$$T(a,b) = 1 \text{ for } S(a,b) > 1.15 \text{ x } max(S(a,b)) - 0.3 \text{ or else}$$

$$T(a,b) = 0$$

Then, the preference matrix is transformed as shown in Table 9.16, but the ranking outcome of distilling the rows and columns would in this case be the same.

Table 9.14 Discordance Matrix

	Size (m^2)	Price ($M)	Location (quality)
House A versus B	1	0	0
House A versus C	0	0	0
House B versus C	0	0.71	0
House B versus A	0	0.86	0
House C versus A	0	0	0
House C versus B	0	0	0

Table 9.15 Preference Matrix

	House A	House B	House C	Row totals
House A		0	0.78	0.78
House B	0.30		0.61	0.91
House C	0.68	0.47		1.15
Column totals	0.98	0.47	1.39	

Table 9.16 Defuzzified Preference Matrix			
	House A	House B	House C	Row totals
House A		0	1	1
House B	0		1	1
House C	1	0		1
Column totals	1	0	2	

9.7.3 Weighting

As the description of MAUT in this chapter hinted, there may be more to weighting than decision-makers realise. We need to distinguish between two principal kinds of weights, in order to avoid a mathematical error. In MAUT, the weights can be called 'trade-off weights' – they represent how much performance against one criterion the decision-maker is willing to sacrifice for a certain amount of performance against another criterion. (The philosophy of doing this is touched upon in Section 9.7.5.) This is not at all the same idea as an 'importance coefficient' as used in ELECTRE. An importance coefficient expresses how important a criterion is compared with the others. Rowley et al. (2012) illustrated this neatly as follows. Imagine a choice between two products, A and B. A costs $6 and will use 40 kWh of electrical power over its life. B is cheaper, costing $4, but it uses more power (60 kWh). If we were using MAUT to compare A and B, with equal weighting, the outcome is A*I*B, a dead heat. What this means is that we are willing to pay $2 for 20 kWh. Of course, we could compare this with the actual price of electricity. If this is not 10 cents/kWh but $1/kWh, and we use zero-max scaling of the two criteria in this MCA, then:

$$100[cents/kWh] = \frac{max(price).w_{price}}{max(energy).w_{energy}} \tag{9.16}$$

and we get $w_{price} = 0.91$ and $w_{energy} = 0.09$ as the actual trade-off weights. Notice that if the maximum price was higher (for example if a new, expensive alternative C were introduced) the trade-off weights would have to change to reflect this information even when the price of electricity is held constant. So trade-off weights say nothing about the actual relative importance of price and energy consumption as issues - they present a quantitative trade-off relationship between indicator scores. On the other hand, importance coefficients tell us how the decision-maker perceives the relative importance of the issues. In the foregoing example, if $w_{price} = 0.91$ were an importance coefficient, we see that the decision-maker perceives cost to be 10 times more important than energy consumption. Clearly, knowing what kind of weight you need for a particular MCA method is critical to asking the right kind of questions when engaging with decision-makers to elicit weights.

9.7.4 Rank Reversal

Ever changed your mind about a major purchase when you heard what someone else bought? You may have experienced rank reversal. Imagine you have the choice of investing in one of two apartments: apartment A costs $1,000,000 and is 110 m^2 in size. B costs $500,000, but it is smaller: 50 m^2. If you weight the criteria equally and compare them in a zero-max-scaled MAUT framework, A*P*B – apartment A is preferred – for twice the price you get more than twice the size. But what happens if you introduce an apartment C into the comparison, either because it is offered to you or your friend bought it? C is enormous compared to the others (300 m^2), but the price tag is not so different to A ($1,100,000). Naturally, if you include it in your MCA, it comes out on top for being such a bargain, but, strangely enough, B*P*A – apartment B is now superior to A! What has happened here is that the scale of the size difference between A and B has been compressed, so the size contributes less to the total utility of A than it did when the MCA had only two alternative options. Note that rank reversal is not confined to MAUT methods but is a common issue across the field of MCA.

Whether this is a problem is a good question, and depends on what values you cherish in decision-making. From a scientific point of view, it may be disturbing that the rank of two options can flip if a third option is introduced into this simple MCA. After all, neither A nor B has changed. On the other hand, many analysts would see this rank reversal as a mathematical phenomenon that is worth embracing because it is consistent with how people actually think and compare alternatives in real life. From this anthropocentric point of view, the phenomenon demonstrates that the MCA approach is a good computational operationalisation of stakeholder thought processes. In any case, this example indicates the importance of correctly identifying genuine alternatives before evaluating criteria and aggregating preferences.

9.7.5 Compensation

As it is with rank reversal, different approaches can be defensible in relation to how MCA handles trade-offs or 'compensation'. The field of MCA may broadly be divided into compensatory and non-compensatory methods. Compensation in this case refers to whether high performance in one category can be used to argue that poor performance in another does not matter. Compensation is a manifestation of e.g. a 'weak sustainability' perspective in which the decision-maker is willing to exchange environmental capital (systems delivering ecosystem services) with other kinds of capital (human or manufactured capital). Such exchanges are not considered to exist under a 'strong sustainability' perspective (Brand, 2009). In an MCA based on strong sustainability, this kind of trade-off is not accepted. For example, Rockström's criteria for a planetary safe operating space are considered independent – violate one criterion and it will not matter how well you do on the others (Rockström et al., 2009 – see Section 2.5.1). However, in many practical situations, we see examples of such trade-offs being made. For example, many affluent people would be happy to eat a poor-quality lunch, if the price of the lunch is very low. In environmental management, we see examples of this attitude – decisions where an endangered habitat is sacrificed for the sake of employment or financial benefits. The job of an analyst supporting a decision-making process is not to dictate what the values of the decision-makers should be, but to make clear what the trade-offs are to the decision-makers and other interested parties and possibly to highlight the values of non-present stakeholders. Furthermore, the analyst needs to make clear whether compensatory methods are aligned with both the type of decision and the values of the particular decision-makers. Understanding the alignment between methods and decision types is a methodological question for experts, but understanding the alignment between a method and the values of the decision-maker is a matter of stakeholder dialogue. Regarding how compensatory different methods are, writers including Guitouni and Martel (1998) and more recently Cinelli et al. (2014) have produced tables indicating whether MCA methods are compensatory, non-compensatory or partially compensatory. PROMETHEE has been put in different categories by different authors, perhaps depending on how the preference functions are implemented – we may consider it to be partially compensatory. Of the other three MCA methods described in this chapter, MAUT and AHP are fully compensatory, while ELECTRE is non-compensatory.

9.8 Dialogue with Stakeholders

This chapter of the book deals with decision-making in situations where there are multiple criteria to consider, and when criteria cannot be expressed using the same metric. Some of them or even many of them may also be qualitative. This is one of many situations where it becomes obvious that decision-making in practice can only partially rely on quantitative numbers and scientific information. However, this does not mean that it cannot be a well-structured and transparent process that can lead to very good decisions that can subsequently be understood by people who were not directly involved.

MCA is not some kind of sausage machine into which we can feed technical data and stakeholder values, and expect durable decisions to pop out the other end. Any expectation that it will function in this way relies on an excessively static, deterministic view of the world. In practice, stakeholder values can vary in the face of information. Perversely, providing information that contradicts a stakeholder's beliefs may sometimes only serve to strengthen those beliefs (Lord et al., 1979). Introduction of a new alternative in a set of alternatives can sometimes create a rank reversal among those already ranked. For reasons like these, the real value of MCA is as a structured, open and pedagogical activity that can interactively guide stakeholders and create consensus via iterative use.

The inclusion of different stakeholders or stakeholder groups in such decision-making processes is in line with the idea of participatory and democratic decision-making as advocated in sustainable development efforts. Having sustainable development as an overarching goal in society also creates a need to address a broader scope in more situations in order to make sure that a holistic approach is adopted, that multiple interests are considered and that decisions do not lead to suboptimal solutions. Therefore, it can be expected that decision-making will increasingly have to handle a wide range of very different criteria and that many different types of stakeholders will be considered and involved in such processes.

9.8.1 A Tale of Urban Water

In practice, engineers and scientists have the opportunity to help set the tone of interactions between the organisations they advise and the general public. For example, in the planning of water infrastructure, approaches vary from strictly expert-driven planning processes in which the public has no say, to processes in which the public elects representatives from beyond the normal sphere of politics to guide development planning. The former approach is suitable for small infrastructure investments that have little impact on the public during construction, and do not change how the public interacts with the operational infrastructure. On the other hand, if the investment does not comply with these two caveats, expert-driven processes can appear to the public as authoritarian exercises in 'listening to DAD' (in which an authority will Decide, Announce and Defend a decision). One example of this approach was the water supply plan for Cochabamba, begun by the Bolivian government under direction from the World Bank. This water supply system renewal and privatisation process included raising prices rapidly and was perceived to affect the rights of local people in several other ways that struck at their livelihoods, and generated fierce opposition (see Figure 9.4). A local coalition of opponents to the process (calling itself the 'Coordinator for the Defence of Water and Life') organised protests. The story of this planning process went on to include rioting in the streets, military-style law enforcement (see Figure 9.5), gunshot injuries and death (Finnegan, 2002), and it came to be known as the 'Cochabamba Water Wars'.

Figure 9.4 Protest art: 'The water is ours!' (photo: James Dryburgh)

Figure 9.5 Street battles in the Cochabamba water wars (photo: Thomas Kruse)

The DAD model has failed in many other places, including rich countries with expectations of reliable infrastructure. In 2005, downward revision of long-term yield calculations for surface water sources and an intense drought had made water supply for Toowoomba (a city atop the Great Dividing Range in Queensland, Australia) an urgent problem for the city council. The council announced its intention to introduce 'indirect potable recycling' into its water supply system – pumping treated recycled water into drinking water dams. It is understandable that the local government felt empowered to make quick decisions when apparently necessary. The process was democratic in the sense that the decision-makers were elected representatives of the people. The city was facing a water supply crisis, requiring someone responsible to act quickly. And for precedents, many riverside cities filter and drink river water downstream from the sewage treatment plants of other cities. You can even consider potable water recycling to be the way of the future – astronauts drink recycled urine (to save weight at lift-off). On the other hand, deliberately redirecting recycled water into drinking water dams is a radical solution for us earthlings, as it goes against hundreds of years of human awareness of the health benefits of keeping toilets and taps apart.

So it was unsurprising that a group calling itself 'Citizens Against Drinking Sewage' sprang into action to fill a perceived information gap. When the national government's funding for the water recycling infrastructure was made conditional on a vote by the local electorate, the council's plan did not succeed. Put simply, the public was not included in the decision-making process until the last minute. The DAD model failed and had the effect of reducing trust in local government (the council). In the end, the public had to instead accept a long, expensive, uphill pipeline from a coastal dam (Wivenhoe). Ironically, plans were made to top up this coastal dam under low-inflow conditions using recycled water obtained in Brisbane. So Toowoomba would have been drinking recycled water anyway, if it were not for the arrival of catastrophically heavy rains in 2008 and the cancellation of that recycling project.

Gold Coast City is a short downhill drive from Toowoomba, and it was also faced with a major water shortage, but the planning process and its outcomes were the exact opposite. In this case, the mayor foresaw the potentially controversial nature of the process, and proposed an inclusive planning process he called Gold Coast Waterfuture. The local government decided to create a citizens 'Advisory Committee', under the guidance of an impartial, external facilitator, to draft the water plan. In addition to some elected council members and council officers, the city invited environmental groups, resident groups, industry groups and state government agencies to provide representatives for the Committee. This was a significant departure from DAD; indeed, the Committee also included high school students to voice the concerns of those who were destined to inherit long-term water supply infrastructure, but who were not normally entitled to vote in government elections. The Committee had the ability to request technical reports on the costs and other impacts of alternative options. In the end, the Gold Coast Waterfuture document included a broad range of initiatives: raising a freshwater dam wall (flooding a forested area), constructing a desalination plant, a dual pipe water recycling scheme, a rainwater tank program and leakage reduction activities. Some of the elements of the plan could easily have generated controversy, but the delivery of the plan in 2005 was remarkably uncontroversial, certainly in comparison to the Toowoomba experience. This indicated the value of public empowerment in the decision-making process. It has been suggested that neither city had a water shortage, but rather a shortage of trust, which the Gold Coast Waterfuture process addressed.

9.9 The Influence of Human Values on Decisions

Human values and cultural worldviews play a key role in forming policy decisions, no less so when they relate to the environment.

Your environmental worldview is your belief system in terms of what you think we can know about the world and how it functions and what our role should be in relation to it. Sometimes, a division is made between three types of environmental worldviews ranging from the more human-centred planetary management view to the more earth-centred environmental wisdom view (Miller & Spoolman, 2012). These can be associated e.g. with the extent to which we feel responsibility towards other living things – see Figure 9.6.

The *planetary management worldview* sees humans as apart from nature and aims to manage nature to provide for the needs and wants of humanity. This approach relies on the belief that we can understand and master nature and that, when allowed to develop freely, technology and human ingenuity will eventually solve problems that

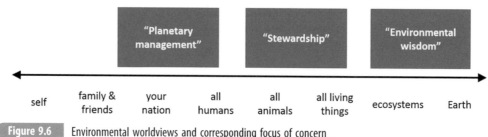

Figure 9.6   Environmental worldviews and corresponding focus of concern

we run into, for example problems like resource scarcity. Therefore, in principle, economic growth is unlimited. With this view, humans are more important than other species and nature exists for our benefit. Success depends on how well we manage to create benefits for people.

In an *environmental stewardship worldview*, we have the ethical responsibility to be caring managers for the Earth. This is a more precautionary and humble approach, and it presumes that since we can run out of resources, we should actively encourage beneficial activities and discourage others. The Earth is a life-support system for us and other organisms that we need to handle with care. Success depends on how well we manage the planet as a life-support system for us and other species.

In an *environmental wisdom worldview*, humans are part of and completely dependent on nature. Nature cannot be mastered, and we should never assume that we can fully understand the complex systems of nature. Further, nature exists not only for humans, but it belongs to all species, and we must find ways to live together on this planet. Success depends on how well we manage to learn how nature maintains and sustains itself and how to integrate this knowledge into our way of life.

If you have an environmental worldview that includes strong notions of environmental wisdom, you will probably think that it is rational to assume that we do not have all the knowledge that we need, and that we must be very careful when we make decisions that can have large impacts on the environment. If you, on the other hand, are tilted towards the planetary management worldview, you may think it is completely unreasonable to hamper economic development by being too precautionary in our decision-making as you are sure we will find ways to solve any problem that will arise. It is easy to understand that the difference between such views can cause political conflicts.

Another conception of how different people imagine the environment belongs to the general framework of 'Cultural Theory'. This has been used to provide different perspectives on risk and environmental assessment based on the idea of the five archetypes shown in Figure 9.7. The figure is based on the idea that your perception of risk is strongly influenced by the degree to which you feel a sense of belonging to a larger group, and the degree to which externally imposed prescriptions ('the grid') control your choices (Thompson et al., 1990).

The hierarchist sees rational environmental control mechanisms as appropriate, and accepts calculated risks where scientific evidence is available. This is the typical view of policy makers and the scientific community. Egalitarians differ in that they have a weaker connection to their grid, and do not accept guidance from external

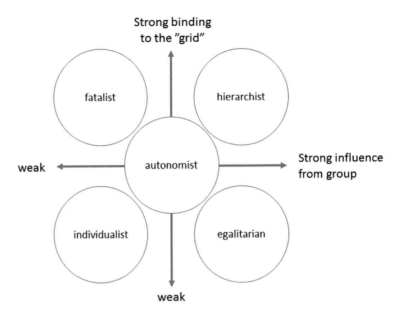

Strong binding
to the "grid"

fatalist   hierarchist

weak ←   autonomist   → Strong influence
                                        from group

individualist   egalitarian

weak

Figure 9.7   Grid-group dependency under Cultural Theory

political or scientific institutions. For people in this group, the precautionary principle and an unwillingness to discount future damages are reflections of their risk aversion. Individualists reject both the grid and group connections, and only worry about risks to which they themselves are clearly exposed. Fatalists see nature as capricious and random, so they have no responsibility for it, while autonomists behave in ways that are not influenced by grid or group affiliations. (For further discussion of factors affecting risk perception and risk aversion, see Section 6.8.)

Of course these are just archetypes, and any real person may exhibit aspects of each. Nevertheless, these archetypes can potentially provide a framework for sensitivity testing the outcomes of MCA processes, and they have been used by many LCA analysts via the Ecoindicator 99 method. Goedkoop and Spriensma (2000) decided to leave fatalists and autonomists out of their work, since these groups are not concerned with trying to manage sustainability, or are not feasible to model. They produced LCIA factors for various impact categories and attempted to generate a method for calculating life cycle impacts as a single indicator via an MCA process. This meant modifying the extent to which different LCIA indicators contributed to the broad damage endpoint categories of 'human health', 'ecosystem quality' and 'resources'. For example, they posit that individualists are not concerned by the depletion of fossil fuels, since they expect to be able to find alternative energy sources that solve at least their personal energy needs (consistent with their planetary management worldview). So, in the Ecoindicator 99 approach, the depletion of fossil fuels does not damage 'resources' (though the depletion of other mineral resources does). In practical terms, this means if you assess a product that consumes fossil fuel, the endpoint indicator for resource impacts does not increase. Furthermore, under this perspective, impacts on the damage category 'human health' are age-weighted, since in the individualist perspective, a person is most valuable between the ages of 20 and 40 years. Neither of

Table 9.17 Weighting Factors in Ecoindicator 99			
Endpoint damage category	Egalitarian	Individualist	Hierarchist
Human health	300	550	400
Ecosystem quality	500	250	400
Resources	200	200	200

these assumptions is used in the egalitarian version of Ecoindicator 99. These are two examples of ways in which the calculations leading to a damage category are modified to reflect cultural worldviews. Furthermore, a set of weights for aggregating damages in the three categories was presented, in which the individualist does not place as much emphasis on the 'world outside' (ecosystem quality) as the other two kinds of decision-maker (Table 9.17). The creators of Ecoindicator did not suggest this information was absolute, but suggested that people using LCA to inform decisions could use the hierarchist model and weights as a default, and check whether the ranking of alternative options remains robust when tested with the egalitarian and individualist value sets.

People with similar values often join forces and create different interest organisations. Greenpeace and Friends of the Earth are examples of non-governmental organisations (NGOs) for people interested in environmental issues. If different stakeholder groups are allowed to make their own valuation and weighting in an MCA process, it might be the case that they come to different conclusions. The principle of NIMBY often appears in discussions around different decisions, illustrating the propinquity factor (see Chapter 6.3). NIMBY is short for 'Not In My Back Yard'. People may, for example, be positive in general to wind power or industrial activities, but when it comes to the location, they do not want to have them close to their own homes. Even when stakeholders are not directly involved in decision-making, it is often valuable to try to imagine what different stakeholder groups would think about the decision at hand. When different stakeholders come to different conclusions on what is the best alternative, a dilemma appears. This dilemma could possibly be solved by modifying the considered decision alternatives and thereby addressing certain concerns. Such a process requires negotiation and can lead to a compromise solution that all groups eventually accept.

Our perception of the professional role of an engineer and what values an engineer should serve depends on our ontological, epistemological and axiological beliefs. *Ontology* is about what really exists – what things are actually real. *Epistemology* concerns what we can actually know about reality. *Axiology* refers to values such as ethics and aesthetics. Beliefs of these kinds vary between people and across regions and have changed over time.

Considering ontological beliefs for example, two different principles are common in engineering and may create tension: the realism principle of 'scientists' and the phenomenological principle of 'designers'. In realism, we can only know the reality that is external to and independent of us and that is governed by absolute 'laws'. In phenomenology, however, the reality is the world we know by interacting with it in an emergent process in which our knowledge changes as we interact with the world (de Figueiredo & da Cunha, 2007).

9.10 Ethics in Practice for the Engineer's Decision-Making Processes

Professional life frequently throws up difficult decisions in which the individual is faced with a choice between serving different valid principles or different genuine stakeholders. When the situation concerns general ethical principles and duties to others, this may be most easily described in qualitative terms, as summarised later in this chapter. When the situation has to do with balancing competing objectives, for example in infrastructure planning, it may be feasible to apply a multicriteria approach such as those already described, but decision-makers will have to be aware of the issue of compensation (see Section 9.7.5).

9.10.1 Managing Ethical Problems and Dilemmas

An ethical dilemma is a complex situation that often involves an apparent mental conflict between moral imperatives, in which to obey one would result in breaking another. For something to qualify as an ethical dilemma, three conditions have to be fulfilled: (1) there must be a decision at hand about which course of action is best, (2) there must be different courses of action to choose from, and (3) no matter what course of action is taken, some ethical principle is compromised. The 'trolley problem' is often used to illustrate a theoretical ethical dilemma. Imagine a tram (or 'trolley') is rushing in an uncontrolled way down a city road. Five people further down the track will be unable to get away in time and will be killed if the trolley hits them. From where you are standing, you can pull a lever and direct the trolley onto a different track, saving the five, but killing another person on this other track. Should you pull the lever? To a utilitarian, pulling the lever is the better option – fewer people die, so this is the best course of action. An alternative viewpoint is that by pulling the lever, you become complicit in a homicide, and an actor in a system that is morally wrong.

We rarely get to steer trolleys, but we often face ethical problems or even dilemmas in our daily life, at least if we actively make an effort to think about the potential consequences of our decisions, e.g. when we decide which products to buy in the shop and when we decide on which information to share with our friends and how. These situations sometimes leave us feeling uneasy and awkward. Ethical dilemmas also appear in the professional roles of engineers. In the classic 'lifeboat problem', a ship's captain in an overcrowded lifeboat has to decide what to do as a storm approaches: throw out a few people to lighten the boat, or watch everybody in it drown. This is a dilemma which is clearly relevant in environmental management, for example when policy makers have to decide how to spend conservation budgets. Dividing a budget across all worthwhile causes may leave too little money or too few resources for each to solve any of them. Other ethical problems or dilemmas may be related to whether one should react to misconduct within a company or among suppliers or collaborators. A fairly recent ethical issue in the field of engineering is how Volkswagen 'cheated' emissions tests for diesel cars. The cheating was discovered outside the company, by a university, so the employees of the company who knew about it apparently chose loyalty to their employer over honesty towards testing authorities and the public. A person who decides to go outside their employer's chain of command and inform outsiders (for example, regulators or even the media) about such misconduct is sometimes referred to as a *whistle-blower*. Cate Jenkins is an example of an environmental whistle-blower. She approached the *New York Times* in 2006 to say that her employer, the USEPA, had lied about the dust hazards firefighters faced when the World Trade Center collapsed in 2001. She eventually won a lawsuit for wrongful dismissal. The most famous whistle-blower today is probably Edward Snowden, who was concerned about government snooping on citizens, and chose to breach the secrecy provisions of his employment contract to tell the world about it. He is currently regarded as a criminal endangering national security by the US government and many of its citizens, while many others regard him as a hero who has helped to protect human rights. Many governments and companies have a whistle-blower policy that ensures that an employee can report misconduct anonymously and without risking anything. In the latter case, misconduct is managed within the company and is reported to upper levels in the organisation. The 'axe problem' is sometimes used to illustrate the dilemma that can appear when you have information that you are obliged to reveal but that may then hurt someone else. The axe is carried by a potential murderer who comes to your house and asks for someone who is inside. If you lie to the potential murderer to protect the person inside your house, you break a moral obligation to always tell the truth. Anything that could happen as a result of your lie, you would then be responsible for, for example if the person you intended to protect has snuck out of the house and meets the potential axe murderer as he leaves the house. On the other hand, if you tell the truth, you could be considered not to have any responsibility for what happens afterwards. This is an extreme version of duty ethics, which is one approach you can take to an ethical dilemma.

It is often useful to apply different ethical principles to an ethical problem or dilemma in decision-making to see whether this can guide your choice of action. This can be referred to as *ethical analysis*. General guidance on ethical matters can be obtained from the Center for Engineering Ethics and Society at the National Academy of Engineering: www.onlineethics.org/. Other frameworks available in literature include the Ethical Cycle (van de Poel & Royakkers, 2011) and the Framework for Solving Ethical Dilemmas (Adawi & Svanström, 2012), both developed in engineering education contexts. Both frameworks rely on a thorough analysis of the dilemma and the application of different ethical principles to it in order to illuminate it from different viewpoints. Firstly, an ethical dilemma (or problem) needs to be understood in terms of who is making the decision and the role and obligations of that person, what the possible options are and what the consequences for different stakeholders of each option are. After that, different ethical principles can be applied to the problem. A set of such principles is briefly described here in the form of questions:

- *What are the potential consequences of the different courses of action?* Consequence (or 'teleological') ethics focusses on the consequences of an action without considering whether the action in itself might be considered right or wrong (i.e. 'the ends justify the means').
- *What are your moral obligations to people or values that might be affected?* Duty (or 'deontological') ethics, on the other hand, focusses on your duties and obligations under the relevant rules and norms (e.g. laws, contracts and prevailing cultural values). You may, for example, have a duty in your profession governed by an ethical code or your company's code of conduct.
- *What are the rights of people (or animals) affected by an action?* Those affected may have certain rights that should be considered.
- *What would it feel like if my choice was made public?* You should consider whether your choice would stand the scrutiny of others (a publicity test). However, remember that there are some decisions of a sensitive nature that are important to take but that would be very difficult to defend should they become public – so this is not to say that all decisions and the reasoning behind them must be made public.
- *What would happen if everyone made this choice in similar situations?* You should also consider whether your choice could be the basis for a universal rule (a universality test).
- *What would it feel like if I and my family were the ones negatively affected?* Finally, you should think about what your choice would be if you or someone close to you were badly affected (an empathy test).

Further guidance on how to resolve an ethical dilemma, or at least to come to a well-thought-out decision, can be found in different ethical codes and principles. For engineers, there are generic ethical codes published by different organisations and there may also be codes of conduct within companies that can provide guidance. Several different codes of ethics for engineers, as published by different organisations, are compiled at www.onlineethics.org/Resources/ethcodes/EnglishCodes.aspx. An example of a corporate code of conduct can be found at: www.basf.com/docu ments/corp/en/about-us/management/BASF_Compliance_Brochure_2012.pdf.

9.11 Review Questions

1. Think of five decisions you will face this weekend. Which of the four types of decisions (see Section 9.2) does each represent? If none of your decisions is an example of one of the types, describe a practical decision that someone else may face which exemplifies one of them.
2. Imagine you were a British voting official in 2016. It has been suggested that the so-called Brexit vote on Britain's exit from the European Union was decided by older people who may be less affected by the downsides of the UK leaving the EU. Imagine that the 'one vote, one value' principle on which normal democracy is based is not sacred, and that each vote shows the voter's age. How could you

theoretically adjust the voting results to take into account that many of those who voted for Brexit will have died of old age during the 2.75-year period between the vote and the end of negotiations (when the changes will be implemented)?

3. Different means for generating alternative options were presented in this chapter (see Section 9.4). Draw a pair of axes: the x-axis represents the degree of structure in the thinking process; the y-axis represents the sense in which the process is backward- or forward-looking. Place each of the seven methods in its appropriate place on the graph.

4. What characteristics do good MCA criteria have? What is the problem with a criterion that does not embody one of these characteristics?

5. You are a sustainability consultant for the minister for water in a dry country. The minister has asked a committee to choose a water recycling system for the national capital city. The committee is primarily concerned about the price and the greenhouse gas emissions of the alternatives, criteria to which they have assigned weights of 60% and 40%, respectively. You observe that the committee members feel the alternatives are pretty similar in price if the difference between alternatives is $100 million or less. Likewise, if the greenhouse gas emissions differ by 1 kg CO_2 equivalent per ML of water, or less. In these cases, they just want the quickest solution you can obtain. On the other hand, if the differences between the systems are $300 million or 2 kg CO_2 equivalent per ML of water, they are interested in why. If they buy a system that emits more than 7 kg CO_2 equivalent per ML of water than an alternative system, your reputation as a sustainability consultant will be shattered, along with your career, and you can expect to lose your family home. If they buy a system that costs $1,500 million more than another alternative and word gets out, public support for the government will evaporate, and you should expect a bloody military coup. Two offers come in: from an American and a Belgian company. System A costs $800 million and emits 15 kg CO_2 equivalent per ML of water, while system B costs $2,000 million and emits 10 kg CO_2 equivalent per ML of water. Use ELECTRE III to recommend an option to the minister.

6. What is the difference between trade-off weights and importance coefficients?

7. What is rank reversal? When does it happen? Does it matter?

8. What are the short-term downsides to public engagement in decision-making? What are the long-term benefits?

9. List some of the key underlying variables used to describe how cultural and environmental worldviews differ between people, and how they may influence their responses in environmental management decision-making.

10. What is an ethical dilemma?

11. Look for information on the Volkswagen emissions testing scandal for some detail on what choices were made that led up to this situation. Select a generic professional role of an engineer that was potentially involved. Apply the framework of ethical analysis provided in this chapter to shed light on a choice that this person made. Would you do the same?

References

ABS (2016) Causes of death, Australia, 2015; 3303.0, Table 1.1. Australian Bureau of Statistics, Canberra. 30 November.

ACC (2011) Cradle-to-gate life cycle inventory of nine plastic resins and four polyurethane precursors. Plastics Division of the American Chemistry Council. http://plastics.americanchemistry.com/LifeCycle-Inventory-of-9-Plastics-Resins-and-4-Polyurethane-Precursors-Rpt-and-App accessed December 2014.

ACS (2018) 12 principles of green engineering. American Chemical Society, www.acs.org accessed February 2018.

AFS (2015) *Hygieniska gränsvärden* [*Exposure Limits*]. AFS 2015: 7.*Arbetsmiljöverkets författningssamling* [The Swedish Work Environment Authority's Statue Collection].

AIHW (2011) Principles on the use of direct age-standardisation in administrative data collections: For measuring the gap between Indigenous and non-Indigenous Australians. Australian Institute of Health and Welfare. Cat. no. CSI 12, www.aihw.gov.au.

Allen DT (2001) Evaluating environmental fate: Approaches based on chemical structure. In DT Allen & DR Shonnard (eds), *Green Engineering: Environmentally Conscious Design of Chemical Processes* (pp. 93–135). Prentice Hall.

Alvarez-Gaitan JP et al. (2013) A hybrid life cycle assessment of water treatment chemicals: An Australian experience. *International Journal of Life Cycle Assessment* 18:1291–1301.

Anastas PT, Warner JC (2000) *Green Chemistry: Theory and Practice*. Oxford University Press.

Anastas PT, Zimmerman JB (2003) Design through the 12 principles of green engineering. *Environmental Science & Technology* 37(5):94A–101A.

Andersson C, Törnberg A, Törnberg P (2014) Societal systems: Complex or worse. *Futures* 63:145–157.

Andrzejewski P, Kasprzyk-Hordern B, Nawrocki J (2005) The hazard of N-nitrosodimethylamine (NDMA) formation during water disinfection with strong oxidants. *Desalination* 176(1–3):37–45.

Antweiler W (2014) *Elements of Environmental Management*. University of Toronto Press.

Arvidsson R, Baumann H, Hildenbrand J (2015) On the scientific justification of the use of working hours, child labour and property rights in social life cycle assessment: Three topical reviews. *International Journal of Life Cycle Assessment* 20 (2):161–173.

Aylward GH, Findlay TJV (1974) *SI Chemical Data*. Second edition. John Wiley & Sons.

Azar C, Holmberg J, Karlsson S (2002) *Decoupling: Past Trends and Prospects for the Future*. EDITA Norstedts Tryckeri AB.

Baumann H, Tillman A-M (2004) *The Hitch Hiker's Guide to LCA*. Studentlitteratur.

Beaulieu L, van Durme G, Arpin M-L, Reveret J-P, Margni M, Fallaha S (2015) *Circular Economy: A Critical Review of Concepts*. CIRAIG.

Benoît C, Mazijn B (eds) (2009) Guidelines for social life cycle assessment of products. www.unep.org/publications/search/pub_details_s.asp?ID=4102 accessed February 2015.

Beychok M (2005) *Fundamentals of Stack Gas Dispersion*. Fourth edition. Self-published.

Birgersson B, Sterner O, Zimerson E (1995) *Kemiska Hälsorisker: toxikologi i kemiskt perspektiv* [*Chemical Health Risks: Toxicology from a Chemical Perspective*]. Liber Ekonomi.

Blasco JL, King A, McKenzie M, Karn M (2017) The KPMG Survey of Corporate Responsibility Reporting. Report 134802. home.kpmg.com accessed March 2018.

Boeker E, van Grondell R (1995) *Environmental Physics*. Wiley.

Boje DM (2001) *Narrative Methods for Organizational and Communication Research*. SAGE Publications.

Boulding K (1956) General systems theory: The skeleton of science. *Management Science* 2(3):197–208.

Box GEP, Draper NR (1987) *Empirical Model-Building and Response Surfaces*. John Wiley & Sons.

Brand F (2009) Critical natural capital revisited: Ecological resilience and sustainable development. *Ecological Economics* 68(3):605–612.

Brans J-P, Mareschal B (2005) PROMETHEE methods. In J Figueira, S Greco & M Ehrgott (eds), *Multiple Criteria Decision Analysis: State of the Art Surveys* (pp. 163–196). Springer.

Brunner PH, Rechberger H (2005) *Practical Handbook of Material Flow Analysis*. https://thecitywasteproject.files.wordpress.com/2013/03/practical_handbook-of-material-flow-analysis.pdf.

Burgman M (2005) *Risks and Decisions for Conservation and Environmental Management*. Cambridge University Press.

Canuckguy and many others: BlankMap-World6.svg; Robert Simmon, NASA: Thermohaline_Circulation_2.png; Robert A. Rohde, minor modifications; Miraceti, derivative work [CC BY-SA 3.0 (https://creativecommons.org/licenses/by-sa/3.0)], via Wikimedia Commons (https://commons.wikimedia.org/wiki/File:Thermohaline_Circulation.svg), 2009.

Carson, R (1962) *Silent Spring*. Houghton Mifflin.

CDC (2015) Injury Mortality Reports 1999 and Onwards (USA). Web-Based Injury Statistics Query and Reporting System / CDC WISQARS. Atlanta: National Center for Injury Prevention and Control, Centers for Disease Control and Prevention / CDC. 22 January.

CDC (n.d.) A U.S. soldier demonstrates the use of DDT-hand spraying equipment. Photo. Working Group at the Centers for Disease Control and Prevention (CDC). https://phil.cdc.gov/details_linked.aspx?pid=2621. www.publicdomainfiles.com/show_file.php?id=13515751817026 accessed May 2018.

Chatwal GR, Powar CB (2007) *Biochemistry*. Global Media.

Chertow MJ (2001) The IPAT Equation and its variants: Changing views of technology and environmental impact. *Journal of Industrial Ecology* 4(4):13–29.

Chu DH, Haake AR, Holbrook K, Loomis CA (2003) The structure and development of skin. In IM Freedberg, AZ Eisen, K Wolff, F Austen, LA Goldsmith & SI Katz (eds), *Fitzpatrick's Dermatology in General Medicine*. Sixth edition. McGraw-Hill.

Churchman CW (1967) Guest editorial: C. West Churchman. *Management Science* 14(4):141–143.

Cinelli M, Coles SR, Kirwan K (2014) Analysis of the potentials of multi criteria decision analysis methods to conduct sustainability assessment. *Ecological Indicators* 46:138–148.

Coleman RA (1986) Placental metabolism and transport of lipid. *Federation Proceedings* 45(10):2519–2523.

Cook J, Oreskes N, Doran PT, Anderegg WRL, Verheggen B et al. (2016) Consensus on consensus: A synthesis of consensus estimates on human-caused global warming, Environ. Res. Lett. 11:048002. doi: 10.1088/1748–9326/11/4/048002

Deubzer O (2011) *E-waste Management in Germany. Deutsche Gesellschaft fuer Internationale Zusammenarbeit (GIZ) and Institute for Sustainability and Peace*. United Nations University.

DHE (2002) *Environmental Health Risk Assessment: Guidelines for Assessing Human Health Risks from Environmental Hazards*. Department of Health and Aging and Health Council, Commonwealth of Australia.

Dierdorff EC, Norton JJ, Drewes DW, Kroustalis CM (2009) Greening of the World of Work: Implications for O*Net-SOC and New and Emerging Occupations. The National Center for O*NET Development. www.onetcenter.org/dl_files/Green.pdf

Doppelt, B (2008) *The Power of Sustainable Thinking*. Earthscan/Routledge.

Earles M, Halog A (2011) Consequential life cycle assessment: A review. *International Journal of Life Cycle Assessment* 16(5): 445–453.

Ekvall T, Tillman A-M (1998) Open-loop recycling: Criteria for allocation procedures. *International Journal of Life Cycle Assessment* 2(3): 155–162.

Ekvall T, Tillman A-M, Molander S (2005) Normative ethics and methodology for life cycle assessment. *Journal of Cleaner Production* 13:1225–1234.

European Chemicals Agency (EChA) (2011) Guidance on Information Requirements and Chemical Safety Assessment. Part A: Introduction to the Guidance Document. European Chemicals Agency. Publication reference ECGA-2011-G-15-EN. [https://echa.europa.eu/documents] accessed September 2018.

European Chemicals Agency (EChA) (2018) Candidate list of substances of very high concern for authorisation. European Chemicals Agency. [https://echa.europa .eu/candidate-list-table] accessed October 2018.

European Economic Community (1975) Council Directive 75/442/EEC of 15 July 1975 on waste.

European Environment Agency (EEA) (2017) Air pollutants by source sector. European Environment Agency. http://ec.europa.eu/eurostat/web/products-euro stat-news/-/EDN-20170602–1?inheritRedirect=true accessed May 2018.

European Environment Agency (2005) EEA core set of indicators – Guide. EEA Technical report No 1/2005.

European Environment Agency (2014) Digest of EEA indicators 2014. EEA Technical report No 8/2014.

European Union (2010) Directive 2010/75/EU of the European Parliament and of the Council of 24 November 2010 on industrial emissions (integrated pollution prevention and control). 2010/75/EU, The European Parliament and the Council of the European Union, *Official Journal of the European Union*, L 334/17.

Fantke P (ed.), Bijster M, Guignard C, Hauschild M, Huijbregts M, Jolliet O, Kounina A, Magaud V, Margni M, McKone TE, Posthuma L, Rosenbaum RK, van de Meent D, van Zelm R (2017) USEtox 2.0 Documentation (Version 1). http://usetox.org accessed May 2018.

Figueiredo AD, Cunha PR (2007) Action research and design in information systems: Two faces of a single coin. In N Kock (ed.), *Information Systems Action Research: An Applied View of Emerging Concepts and Methods*. Springer.

Fimreite N (1970) Mercury uses in Canada and their possible hazards as sources of mercury contamination. *Environmental Pollution* 1(2):119–131

Finnegan W (2002) Leasing the rain. *New Yorker*, 8 April. http://web.archive.org/web/20070929151555/http://www.waterobservatory.org/library.cfm?refID=33711 accessed June 2016.

Finnveden G, Moberg Å (2005) Environmental systems analysis tools: An overview. *Journal of Cleaner Production* 13(12):1165–1173.

Fried R (2014) Europe passes historic law: Big corporations must report on sustainability. *Daily Sustainable Business News*, 15 April. http://lcrn.org.uk/europe-passes-historic-law-big-corporations-must-report-sustainability/ accessed May 2018.

Frosch RA, Gallopoulos NE (1989) Strategies for manufacturing. *Scientific American* 189(3):152.

Fuller S, Petersen S (1996) *Life-Cycle Costing Manual*. Handbook 135. National Institute for Standards and Technology.

Goedkoop M, Heijungs R, Huijbegts M, de Schryver A, Struijs J, van Zelm R (2008) *ReCiPe 2008: A Life Cycle Impact Assessment Method Which Comprises Harmonised Category Indicators at the Midpoint and the Endpoint Level*. VROM.

Goedkoop M, Spriensma R (2000) The Eco-indicator 99: A damage oriented method for life cycle impact assessment – Methodology report. PRé Consultants B. V. www.pre.nl accessed December 2016.

Goldenberg J, Mazursky D, Solomon S (1999) Creative sparks. *Science* 285 (5433):1495–1496.

Govindan K, Brandt Jepsen M (2015) ELECTRE: A comprehensive literature review on methodologies and applications. *European Journal of Operational Research* 250 (1):1–25.

Graedel, T (1994) Industrial ecology: Definition and implementation. In R Socolov, C Andrews, F Berkhout & V Thomas (eds), *Industrial Ecology and Global Change*. Cambridge University Press

Guinée J (ed.) (2002) *Handbook on Life Cycle Assessment*. Springer.

Guitouni A, Martel J-M (1998) Tentative guidelines to help choosing an appropriate MCDA method. *European Journal of Operational Research* 109:501–521.

Haas C, Crockett C, Rose J, Gerba C, Fazil A (1996) Assessing the risk posed by oocysts in drinking water. *Journal of American Water Works Association* 88:131–136.

Hanrahan G (2012) *Key Concepts in Environmental Chemistry.* Academic Press/ Elsevier.

Hardin C, Knopp J (2013) *Biochemistry: Essential Concepts.* Oxford University Press.

Harjula H (1998) Extended producer responsibility – Phase 2 – Case study on the Dutch Packaging Covenant. ENV/EPOC/PPC(97)22/REV2. Environmental Policy Committee, Environment Directorate, Organisation for Economic Co-Operation and Development. www.oecd.org/officialdocuments/publicdisplaydocu mentpdf/?doclanguage=en&cote=env/epoc/ppc(97)22/rev2

Harte J (1988) *Consider a Spherical Cow.* University Science Books.

Hauschild MZ, Huijbregts M, Jolliet O, Macleod M, Margni M, van de Meent D, Rosenbaum RK, McKone TE (2008) Building a model based on scientific consensus for life cycle impact assessment of chemicals: The search for harmony and parsimony. *Environmental Science and Technology* 42:7032–7037.

Hawkins TR, Matthews DH (2009) A classroom simulation to teach economic input-output life cycle assessment. *Journal of Industrial Ecology* 13 (4):622–637.

Heijungs R, Guinée J, Huppes G, Lankreijer RM, Udo de Haes HA, WegenerSleeswijk A, Ansems AMM, Eggels PG, van Duin R, de Goede HP (1992) *Environmental Life Cycle Assessment of Products.* Guide. Report No. 9266, CML, Leiden University.

Helander HF, Fändriks L (2014) Surface area of the digestive tract – revisited. *Scandinavian Journal of Gastroenterology* 49(6):681–689.

Helfield JM, Naiman RJ (2001) Effects of salmon-derived nitrogen on riparian forest growth and implications for stream productivity. *Ecology* 82(9):2403–2409.

Hillson D (2009) Swine flu and risk perception. *Project Management Today* June.

Hites RA (2007). *Elements of Environmental Chemistry.* John Wiley and Sons.

Holmberg J, Robert K-H, Eriksson K-E (1995) Socio-ecological principles for a sustainable society. In J Holmberg, Socio-Ecological Principles and Indicators for Sustainability. Draft doctoral thesis paper, Chalmers University of Technology. Published in R Costanza, S Olman & J Martinez-Alier (eds) (1996) *Getting Down to Earth: Practical Applications of Ecological Economics.* Island Press.

HOV (2018) Effekter av kalkning på bottenfaunan I rinnande vatten [Effects of liming on bottom fauna in running water]. Havs och Vattemyndigheten [Swedish Agency for Marine and Water Management]. Report 2018: 4.

ICEX (2009) Desalination in Spain. *MIT Technology Review.* http://icex.technolo gyreview.com/articles/2009/01/desalination-in-spain/ accessed May 2018.

IPCC (2014a) Climate change 2014: Mitigation of climate change. Contribution of Working Group III to the Fifth Assessment Report of the Intergovernmental Panel on Climate Change. In O Edenhofer, R Pichs-Madruga, Y Sokona, E Farahani, S Kadner, K Seyboth, A Adler, I Baum, S Brunner, P Eickemeier, B Kriemann, J Savolainen, S Schlömer, C von Stechow, T Zwickel & JC Minx (eds), Cambridge University Press.

IPCC (2014b) Climate change 2014: Synthesis report. Contribution of Working Groups I, II and III to the Fifth Assessment Report of the Intergovernmental Panel on Climate Change. In Core Writing Team, RK Pachauri & LA Meyer (eds.), IPCC.

Ishizaka A, Nemery P (2013) *Multi-Criteria Decision Analysis: Methods and Software*. Wiley.

ISO (2015a) *ISO 9001: 2015 Quality Management Systems: Requirements.* International Standardization Organization.

ISO (2015b) *ISO 14001: 2015 Environmental Management Systems: Requirements with Guidance for Use.* International Standardization Organization.

Jaworski NA, Howarth RW, Hetling JL (1997) Atmospheric deposition of nitrogen oxides onto the landscape contributes to coastal eutrophication in the northeast United States. *Environmental Science and Technology* 31(7): 1995–2004.

Kitzes J (2013) An introduction to environmentally-extended input-output analysis. *Resources* 2:489–503.

Klaassen C (2013) *Casarett and Doull's Toxicology: The Basic Science of Poisons.* Eighth edition. McGraw Hill.

Kloepffer W (2009) Life cycle sustainability assessment of products. *International Journal of Life Cycle Assessment* 13(2):89–95.

Knowles JR (1980) Enzyme-catalyzed phosphoryl transfer reactions. *Annual Review of Biochemistry* 49: 877–919.

Kobayashi Y, Peters G, Ashbolt N, Khan S (2015) Assessing burden of disease as disability adjusted life years in life cycle assessment. *Science of the Total Environment* 530–531:120–128.

Kudo A, Fujikawa Y, Miyahara S, Zheng J, Takigami H, Sugahara M, Muramatsu T (1998) Lessons from Minamata mercury pollution, Japan: After a continuous 22 years of observation. *Water Science and Technology* 38(7):187–193.

Kuntz E (2008) *Hepatology: Textbook and Atlas.* Springer.

Kwok ESC, Atkinson R (1995) Estimation of hydroxyl radical reaction rate constants for gas-phase organic compounds using a structure-reactivity relationship: An update. *Atmospheric Research* 29(14):1685–1895.

Li H-F, Wang J-J (2007) An improved ranking method for ELECTRE III. International Conference on Wireless Communications, Networking and Mobile Computing, WiCOM 2007, pp. 6659–6662.

Lidsky TI, Scheider JS (2003) Lead neurotoxicity in children: Basic mechanisms and clinical correlates. *Brain* 126:5–19.

Liu X, Guo Z, Roache NF, Mocka CA, Allen MR, Mason MA (2015) Henry's Law constant and overall mass transfer coefficient for formaldehyde emission from small water pools under simulated indoor environmental conditions. *Environmental Science and Technology* 49:1603–1610.

Lord CG, Ross L, Lepper MR (1979) Biased assimilation and attitude polarisation: The effects of prior theories on subsequently considered evidence. *Journal of Personality and Social Psychology* 37:2098–2109.

Lundie S, Ashbolt N, Livingston D, Lai E, Kärrman E, Blaikie J, Anderson J (2008) Sustainability Framework. WSAA Occasional Paper No. 17. February. Water Services Association of Australia.

Masters GM, Ela WP (2008) *Introduction to Environmental Engineering and Science.* Third edition. Pearson.

McDonough W, Braungart M (2002) *Cradle to Cradle: Remaking the Way We Make Things.* North Point Press.

Mead MN (2008) Contaminants in human milk: Weighing the risks against the benefits of breastfeeding. *Environmental Health Perspectives* 116(10):A426–A434.

Meylan WM, Howard PH (1995) Atom/fragment contribution method for estimating octanol-water partition coefficients. *Journal of Pharmaceutical Sciences* 84 (1):83–92.

Mihelcic JR, Zimmerman JB (2010) *Environmental Engineering: Fundamentals, Sustainability, Design*. John Wiley and Sons.

Millennium Ecosystem Assessment (2005) *Ecosystems and Human Well-Being: Synthesis*. Island Press.

Miller GT, Jr, Spoolman SE (2012) *Living in the Environment*. 17th international edition. Cengage Learning.

Mitchell C, White S (2003) Forecasting and backcasting for sustainable urban water futures. *Water: Journal of the Australian Water Association* 30(5):25–30.

Morgan RF (2012) Environmental impact assessment: The state of the art. *Impact Assessment and Project Appraisal* 30(1):5–14.

Munier N (2011) *A Strategy for Using Multicritera Analysis in Decision-Making*. Springer.

NASA (2008) Thermohaline_Circulation. https://commons.wikimedia.org/wiki/ File:Thermohaline_Circulation_2.png or Wikimedia commons: https://commons .wikimedia.org/wiki/File:Thermohaline_Circulation.svg.

NASA (2017) What is Earth's energy budget? Five questions with a guy who knows. www.nasa.gov/feature/langley/what-is-earth-s-energy-budget-five-questions-with-a-guy-who-knows accessed 14 April 2018.

National Weather Service (n.d.) Layers of the atmosphere. www.weather.gov/jet stream/layers accessed 14 April 2018.

NHMRC, NRMMC (2011) Australian Drinking Water Guidelines Paper 6 National Water Quality Management Strategy. National Health and Medical Research Council, National Resource Management Ministerial Council, Commonwealth of Australia, Canberra.

Nilsson J (2003) Introduktion till riskanalysmetoder [Introduction to risk analytical methods]. Report 3124. Department of Fire Safety Engineering, Lund University, Sweden.

NOAA (2018) National Oceanic and Atmospheric Administration, Earth System Research Laboratory, Global Monitoring Division www.esrl.noaa.gov/gmd/ccgg/ trends/full.html.

Oberbacher B, Hansjörg N, Klöpffer W (1996) LCA: How it came about: An early systems analysis of packaging for liquids. *International Journal of Life Cycle Assessment* 1(2):62–65.

Ortwein BM (2003) The Swedish legal system: An introduction. *Indiana International and Comparative Law Review* 13(2):405–455.

Peters GM (2009). Popularize or publish: Growth in Australia. *International Journal of Life Cycle Assessment* 14(6):503–507.

Peters GM, Blackburn NJ, Armedion M (2013) Environmental assessment of air to water machines: Triangulation in the presence of scope uncertainty. *International Journal of Life Cycle Assessment* 18:1149–1157.

Peters G, Granberg H, Sweet S (2015) The role of science and technology in sustainable fashion. In K Fletcher & M Tham (eds), *The Handbook of Sustainable Fashion* (chapter 18). Routledge.

Reynolds HY (1985) Phagocytic defense in the lung. *Antibiotics and Chemotherapy* 36:74–87.

Ritthoff M, Rohn H, Liedtke C (2002) *Calculating MIPS: Resource Productivity of Products and Services*. Wuppertal Spezial 27e. Wuppertal Institute for Climate, Environment and Energy.

Robèrt K-H, Daly H, Hawken P, Holmberg J (1997) A compass for sustainable development. *International Journal of Sustainable Development & World Ecology* 4(2):79–92.

Rockström J, Steffen W, Noone K, Persson Å, Chapin S, Lambin E et al. (2009). Planetary boundaries: Exploring the safe operating space for humanity. *Ecology and Society* 14(2):32. www.ecologyandsociety.org/vol14/iss2/art32/ accessed December 2014.

Rowley HV, Peters GM, Lundie S, Moore SJ (2012) Aggregating sustainability indicators: Beyond the weighted sum. *Journal of Environmental Management* 111:24–33.

Roy B (1968) Classement et choix en présence de points de vue multiples (la méthode ELECTRE) [Classification and choice in the presence of multiple points of view (the ELECTRE method)]. *Revue d'Informatique et de Recherche Opérationnelle* 2(8):57–72.

Roy B (1981) The optimisation problem formulation: Criticism and overstepping. *Journal of the Operational Research Society* 32(6):427–436.

Roy B (1990) The outranking approach and the foundations of ELECTRE methods. In CA Bana e Costa (ed.), *Readings in Multiple Criteria Decision Aid*. Springer.

de Saint-Exupéry A (1939) *Terre des Hommes*. First edition. Gallimard.

Sandin G, Clancy G, Heimersson S, Peters G, Svanström M (2014) Making the most of LCA in technical inter-organisational R&D projects. *Journal of Cleaner Production* 70:97–104.

Sandin G, Peters GM, Pilgård A, Svanström M, Westin M (2011) Integrating sustainability consideration into product development: A practical tool for identifying critical social sustainability indicators and experiences from real case application. In M Finkbeiner (ed.), *Towards Life Cycle Sustainability Management* (pp. 3–14). Springer.

Schumacher EF (1973) *Small Is Beautiful: A Study of Economics as if People Mattered*. HarperCollins.

Sherwood SC, Huber M (2010) An adaptability limit to climate change due to heat stress. *Proceedings of the National Academy of Sciences* 107(21):9552–9555.

Smeets E, Weterings R (1999) Environmental indicators: Typology and overview. Technical report No 25, European Environment Agency.

Solomon S, Ivy DJ, Kinnison D, Mills MJ, Neely RR, III, Schmidt A (2016) Emergence of healing in the Antarctic ozone layer. *Science* 353(6296):269–274.

Sroufe R (2003) Effects of environmental management systems on environmental management practices and operations. *Production and Operations Management* 12(3):416–431.

Steen B (2016) Calculation of monetary values of environmental impacts from emissions and resource use: The case of using the EPS 2015d Impact Assessment Method. *Journal of Sustainable Development* 9(6):15–33.

Steffen W, Richardson K, Rockström J, et al. (2015) Planetary boundaries: Guiding human development on a changing planet. *Science* 347(6223): 736–746.

Svanström M, Adawi T (2012) Dealing with dilemmas: Designing a compulsory PhD course on research ethics and sustainable development. 9th ICED Conference. 23–25 July 2012. Bangkok. Thailand.

Sveriges Riksbank (2011) Ny sedel – och myntserie Format, material och färger [New note and coin series – format, material and colours]. Swedish Riksbank (the Swedish Central Bank). Dnr 2008–286-ADM. http://archive.riksbank.se/ Pagefolders/59191/Rapport%20Format%20material%20och%20f%C3%A4rger %20.pdf accessed October 2016.

Swarr TE, Hunkeler D, Klöpffer W, Pesonen H-L, Ciroth A, Brent AC, Pagan R (2011) *Environmental Life Cycle Costing: A Code of Practice.* Society of Environmental Toxicology and Chemistry (SETAC).

Tessier A, Campbell PGC, Bisson M (1979) Sequential extraction procedure for the speciation of trace metals. *Analytical Chemistry* 51(7):844–851.

Thoma G, Swofford J, Popov V, Soerens T (1999) Effect of dynamic competitive sorption on the transport of volatile organic chemicals through dry porous media. *Water Resources Research* 35(5):1347–1359.

Thompson M, Ellis R, Wildavsky A (1990) *Cultural Theory.* Westview Print.

TNS (2018) The Natural Step Framework. www.thenaturalstep.org/our-approach/ accessed February 2018.

Tong M, Neusner A, Longato L, Lawton M, Wands JR, de la Monte SM (2009) Nitrosamine exposure causes insulin resistance diseases: Relevance to type 2 diabetes mellitus, non-alcoholic steatohepatitis, and Alzheimer's disease. *Journal of Alzheimer's Disease* 17:827–844.

Tortajada C (2006) Water Transfer from the Ebro River. Case study report for the 2006 Human Development Report. United Nations Development Programme. hdr .undp.org/en/content/water-transfer-ebro-river [accessed February 2017]

Turner DB (1994) *Workbook of Atmospheric Dispersion Estimates: An Introduction to Dispersion Modeling.* Second edition. CRC Press.

UNECE (2017) Emission data tables. United Nations Economic Commission for Europe. www.unece.org/env/lrtap/status/lrtap_s.html accessed February 2017.

UNEP (2015) Annual Report 2014. United Nations Environment Programme. www .unep.org/publications/ accessed February 2017.

USDDHS (2010) Biosafety in Microbiological and Biomedical Laboratories. *Fifth edition.* US Department of Health and Human Services. HHS Publication No. (CDC) 21–1112.

US Department of the Interior (2017). The water cycle. Howard Perlman, USGS, John Evans. US Geological Survey. http://water.usgs.gov/edu/watercycle.html accessed 23 April 2018.

USDA (2012). Microbial Risk Assessment Guideline: Pathogenic Organisms with Focus on Food and Water. US Department of Agriculture/Food Safety and

Inspection Service and US Environmental Protection Agency. FSIS Publication No. USDA/FSIS/2012–001; EPA Publication No. EPA/100/J12/001.

USEPA (2011) *Exposure Factors Handbook: 2011 Edition*. National Center for Environmental Assessment. EPA/600/R-09/052F. United States Environmental Protection Agency Available from the National Technical Information Service, Springfield, VA, and online at www.epa.gov/expobox/exposure-factors-handbook-2011-edition.

USEPA (2017) Basic information about the Integrated Risk Information System. United States Environmental Protection Agency. www.epa.gov/iris/basic-informa tion-about-integrated-risk-information-system accessed January 2018.

USEPA (2018) Effects of acid rain. United States Environmental Protection Agency www.epa.gov/acidrain/effects-acid-rain accessed May 2018.

van de Poel I, Royakkers L (2011) *Ethics, Technology, and Engineering: An Introduction*. Wiley-Blackwell.

Walker CH, Hopkin SP, Sibly RM, Peakall DB (2006) *Principles of Ecotoxicology*. Third edition. (Chapters 2 and 3 are especially relevant to Chapter 4 in this book.)

Wenzel H, Alting L (1999) Danish experience with the EDIP tool for environmental design of industrial products. In *Proceedings: 1st International Symposium on Environmentally Conscious Design and Inverse Manufacturing, EcoDesign 1999*, pp. 370–379.

WHO (2004) 10 things you need to know about DDT use under the Stockholm Convention. World Health Organization. WHO/HTM/RBM/2004.55 apps.who. int/iris/bitstream/10665/68633/1/WHO_HTM_RBM_2004.55.pdf accessed February 2017.

WHO (2013) WHO methods and data sources for global burden of disease estimates 2000–2011. World Health Organization. Global Health Estimates Technical Paper WHO/HIS/HSI/GHE/2013.4.

WHO (2016) Inter-country comparison of mortality for selected cause of death: Gun death from undetermined cause in the United Kingdom. European Detailed Mortality Database (DMDB). Copenhagen: World Health Organization Regional Office for Europe. 22 June

WICEE (2014) Material intensity of materials, fuels, transport services. Wuppertal Institute for Climate, Environment and Energy. https://wupperinst.org/uploads/ tx_wupperinst/MIT_2014.pdf.

Wiedema BP, Frees N, Nielsen A-M (1999) Marginal production technologies for life cycle inventories. *International Journal of Life Cycle Assessment* 4:48–56.

Winterton N (2001) Twelve more green chemistry principles. *Green Chemistry* 3(6): G73–G75.

Wirgin I, Roy NK, Loftus M, Chambers RC, Franks DG, Hahn ME (2011) Mechanistic basis of resistance to PCBs in Atlantic tomcod from the Hudson River. *Science* 331(6022):1322–1325.

WMO (2002) World Meteorological Organization, Scientific Assessment of Ozone Depletion: 2002, Global Ozone Research and Monitoring Project, WMO Report 47, Geneva.

Wolf J (1947). Disney DDT wallpaper advert. Photo. www.flickr.com/photos/joe behr/6214943496 accessed May 2018.

WWF (2016) *Living Planet Report 2016: Risk and Resilience in a New Era.* WWF International.

Zarghami M, Szidarovszky F (2011) *Multicriteria Analysis Applications to Water and Environment Management.* Springer, p. 134.

Index

abiotic
    factor, 33
    resource, 25
    transformation, 76
acceptable risk, 132, 133
acenaphthene, 127, 128
acetaldehyde, 135
acetyl coenzyme A, 110
acetylcholine, 112
acid rain, 1, 45
acidification, 45
    ocean, 48
    soil, 47–48
acidifying substances, 45
active transport, 97
adenosine diphosphate, 109
adenosine triphosphate, 109
advection, 73
agonist, 113
AHP. *See* analytic hierarchy process
albedo, 20, 55
allergen, 114
allergic reaction, 114
allocation
    50/50, 155
    by partitioning, 152
    by quality, 155
    by system expansion, 152
    closed-loop approximation, 155
    cut-off, 153
    disposal loading, 155
    economic, 153
    extraction loading, 154
    for open-loop recycling, 153
    in ISO14044, 152
    physiochemical, 152
Alzheimer's disease, 115
analytic hierarchy process, 203, 216
analytical solution, 82, 87, 89
aniline, 99
Annex XIV, under REACH, 178
Annex XVII, under REACH, 179
antagonist, 116
antagonistic effect, 192
Anthropocene period, 19, 39, 58
antibody, 114
antigen, 113
applied scientist, 1
appropriate technology, 14

areas of protection in LCIA, 160
arsenic, 62
asbestos, 106
Asian brown cloud, 52
atmosphere, 20, 67
ATP. *See* adenosine triphosphate
attributional LCA, 146, 155
authorisation, under REACH, 178

B cell, 114
backcasting, 200
Baltic Sea, 49
Basel Convention, 176
BAT. *See* Best Available Technology
BATNEEC. *See* Best Available Techniques Not
    Entailing Excessive Costs
belief system, 219
benzo(a)pyrene, 114
benzoic acid, 99
Best Available Technology, 14
Best Practicable Environmental Option, 14
beta-Poisson function, 131
bile, 107
bile duct, 107
bioaccumulation, 59, 60, 68, 70
bioavailability, 68, 75
biobased economy, 7–8
biocapacity, 39
bioconcentration, 60, 68
biodiversity, 24, 34, 39, 40
biogeochemical cycles, 28
biological membrane, 95
biomagnification, 60, 68
biorefinery, 7
biosafety levels, 137
biosphere, 27–28, 67
biotic transformation, 75
biotransformation, 75, 103
bioturbation, 75, 76
blood, 100, 108
blood-brain barrier, 103
bottom-up control, 32
Bowman's capsule, 108
BPEO. *See* Best Practicable Environmental Option
brainstorming, 200
    negative, 200
    with SWOT, 200
Brownian motion, 95
Brundtland Commission, 2